Exercise Physiology

Edited by Ricardo Ferraz, Henrique Neiva,
Daniel A. Marinho, José E. Teixeira,
Pedro Forte and Luís Branquinho

Published in London, United Kingdom

IntechOpen

Supporting open minds since 2005

Exercise Physiology
http://dx.doi.org/10.5772/intechopen.95586
Edited by Ricardo Ferraz, Henrique Neiva, Daniel A. Marinho, José E. Teixeira, Pedro Forte and Luís Branquinho

Contributors
Júlio Costa, Pedro Figueiredo, João Brito, Fábio Y. Nakamura, Mikhail Shestakov, Anna Zubkova, Natàlia Balagué, Sergi Garcia-Retortillo, Robert Hristovski, Plamen Ch. Ivanov, José E. Teixeira, Pedro Forte, Ricardo Ferraz, Luis Branquinho, Antonio J. Silva, Tiago M. Barbosa, António M. Monteiro, Luis Branquinho, Henrique P. Neiva, Mario C. Marques, Daniel A. Almeida Marinho, Thai Duy Nguyen, Mohamed Magdy Aly Hassan ElMeligie, Vaclav Bunc, Erich Hohenauer, Sheldon Magder, Ana Pereira, Ana Cristina Corrêa Figueira, Luís Leitão, José Alberto Duarte, Rita Ferreira, Kotaro Suzuki

Notice
Statements and opinions expressed in the chapters are these of the individual contributors and not necessarily those of the editors or publisher. No responsibility is accepted for the accuracy of information contained in the published chapters. The publisher assumes no responsibility for any damage or injury to persons or property arising out of the use of any materials, instructions, methods or ideas contained in the book.

First published in London, United Kingdom, 2022 by IntechOpen
IntechOpen is the global imprint of INTECHOPEN LIMITED, registered in England and Wales, registration number: 11086078, 5 Princes Gate Court, London, SW7 2QJ, United Kingdom
Printed in Croatia

British Library Cataloguing-in-Publication Data
A catalogue record for this book is available from the British Library

Additional hard and PDF copies can be obtained from orders@intechopen.com

Exercise Physiology
Edited by Ricardo Ferraz, Henrique Neiva, Daniel A. Marinho, José E. Teixeira, Pedro Forte and Luís Branquinho
p. cm.
Print ISBN 978-1-83969-951-1
Online ISBN 978-1-83969-952-8
eBook (PDF) ISBN 978-1-83969-953-5

Meet the editors

Ricardo Ferraz obtained his BSc in Sports and Physical Education from the Faculty of Sports of the University of Porto (Portugal) in 2006 and his Ph.D. in Sport Sciences from the University of Beira Interior, Portugal, in 2016. He is a professor at the University of Beira Interior, teaching graduate, master's, and doctoral students, and a member of the Research Centre for Sports Sciences, Health Sciences and Human Development (CIDESD). His main research interests are sports physiology, psychophysiology, training control, physical education and performance evaluation in team sports. His more than 50 published works include books, book chapters, and papers in peer-reviewed journals and conference proceedings.

Henrique Neiva obtained his Ph.D. in Sport Sciences at the University of Beira Interior in 2015, supported by a grant from the Science and Technology Foundation. His thesis, entitled "The effect of warm-up on swimming performance: the impact of volume, intensity and post-warm-up recovery in elite swimmers", was awarded the grade "Excellent". He has more than 50 scientific publications to his name and has peer-reviewed several papers in high-ranked journals. A former international swimmer and Level 3 coach, his original academic interest was in training and competition, specifically in swimming. His interest in exercise and health, and specifically in technological development, has emerged in the last two years, mainly because of recent post-doctoral research associated with the NanoSTIMA project in UTAD/CIDESD. He has been a member of the Research Centre for Sports Sciences, Health Sciences and Human Development since 2015.

José Eduardo de Araújo Teixeira graduated in Cardiopneumology from the Castelo Branco Polytechnic Institute in 2013 and in Sports from the Polytechnic Institute of Bragança in 2017. He has a Ph.D. in Sports Sciences from the University of Trás-os-Montes and a Master's in Exercise and Health from the Polytechnic Institute of Bragança, both awarded in 2022. He has published a number of studies on aspects of performance analysis such as exercise physiology, monitoring training load, and tactical analysis. José Eduardo's fields of interest extend to medical and health sciences, focusing on cardiovascular, respiratory and metabolic systems. He is a research fellow at the Centre for Physical Condition Evaluation and Exercise Prescription (CAPE) at the Polytechnic Institute of Bragança, Portugal.

Pedro Miguel Gomes Forte obtained a degree in Sports in 2012 and a Master's in Exercise and Health in 2014, both from the Polytechnic Institute of Bragança. He completed his Ph.D. in Sports Sciences at the University of Beira Interior in 2018. He is Coordinator Professor at the Instituto Superior de Ciências Educativas do Douro and a Guest Professor at the Instituto Politécnico de Bragança. He is also a basketball and football coach. The focus of his sports science research is biomechanics, and his particular interests are prevalence, posture, youth, asymmetries, risk factors, football, training, analysis, performance, CFD, aerodynamics, drag, hydrodynamics, and kinematics.

Luís Branquinho's Ph.D. in Sports Science from the University of Beira Interior was awarded the top grade, "Approved with Distinction". He is currently both a professor researching at the Higher Institute of Educational Sciences of the Douro and a football coach licensed and recognised by UEFA. He has published several scientific articles and books in the field of sports science and his area of expertise is football training.

Since 2008 Daniel Almeida Marinho has been an associate professor at UBI, where he has been the Director of the Ph.D. Sport Sciences course since 2013. He obtained a degree in Sport and Physical Education at UP in 2004, a Ph.D. in Sport Sciences at UTAD in 2009, and Habilitation in Sport Sciences at UBI in 2013. He has supervised 20 graduate theses, 20 MD theses, and ten Ph.D. theses, and has published over 150 academic titles, including ten books and book chapters, 60 articles in SCI journals, 70 articles in peer-reviewed indexed international journals, and ten articles in peer-reviewed indexed national journals. He has participated in the organisation of 15 national and international scientific meetings and has been a member of several Ph.D. and MD juries. Since 2013 he has been working with the Portuguese Swimming Federation, where he is responsible for coaching courses and coordinates the Evaluation and Control Training Centre.

Contents

Preface

Exercise physiology is one of the most researched sports sciences, with extensive practical applications ranging from sports performance to health and well-being. *Exercise Physiology* brings together emerging research in this area, as well as exploring future perspectives. The book is divided into three fundamental sections: (i) Exercise, Brain and Cognitive Functioning; (ii) Energy and Human Performance; and (iii) Physical Exercise and Immune Response.

The first section, "Exercise, Brain and Cognitive Functioning" links exercise physiology to the brain and cognitive functioning.

The chapter "The Performance during the Exercise: Legitimizing the Psychophysiological Approach" looks at the existing evidence and new perspectives on the role of the brain as a central regulator of performance, questioning the complex interdependencies and interrelations between fatigue and physical exercise from a psychophysiological perspective, and bringing together concepts such as pacing behavior, decision-making, self-regulation of effort, prior knowledge of the duration of the task, and perception of effort. Sleep is extremely important for numerous biological functions, and sleep deprivation can have significant effects on athletic performance in the short, medium, and long term.

The chapter "The Importance of Sleep in Athletes" explores sleep as an essential component in athletes' recovery from fatigue, mainly due to its restorative physiological and psychological effects. Monitoring athletes' sleep patterns can optimize recovery strategies, health, performance, fitness, and fatigue status. The authors discuss the following training- and competition-related factors: 1) sleep patterns and disorders; (2) sleep and optimal functioning; (3) screening, tracking, and assessment of sleep; and (4) sleep interventions (i.e., sleep hygiene).

The chapter "Physiological Adaptions to Acute Hypoxia" describes the main types of hypoxia in the oxygen cascade, as well as the physiological consequences of acute hypoxia for training and health. Tissue hypoxia can be caused by any obstacle in the oxygen cascade, and by hypoxaemia, anaemia, stagnation, and histotoxic hypoxia.

"Potential of Physical Activity-Based Intervention on Sleep in Children with and without Autism Spectrum Disorder" also addresses the topic of sleep quality, but from a more clinical perspective, reporting on sleep disorders in children with autism spectrum disorder (ASD). Physical activity is described in the literature as an optimal non-pharmacological approach for improving sleep. The research presented in this chapter studied 75 children, 57 of whom had ASD, with 18 who had typical development as a control. The subjects wore an accelerometer monitor (Sense Wear® Pro Armband 3, Body media) for six consecutive days and nights to assess their sleep and physical activity. Moderate to vigorous physical activity has been shown to be effective in improving sleep in children both with and without ASD. The chapter concludes with recommendations for improving sleep quality and reducing sedentary behavior in both children with ASD and children in general. Physical inactivity and stress at work have been linked to the sedentary lifestyles and lack of time commonly observed today. With market demand for bioactive compounds that mimic or potentiate the effects of exercise, exercise mimetics has emerged as a controversial topic in sports and exercise physiology. While this idea is attractive for those seeking quick results while avoiding the need for a lot of exercise, it is controversial due to the complexity of the molecular and physiological mechanisms involved.

The chapter "Exercise Mimetics: An Emerging and Controversial Topic in Sport and Exercise Physiology" explains the challenges of designing a pill that can reliably provide the myriad and complex adaptations afforded by exercise training, with a focus on skeletal muscle. The multifaceted human physiological response to exercise, with its inherent redundancy, may mean weight-loss pills are no more than a utopia. Moreover, exercise promoting human health should adhere to evidence-based guidelines and pharmaceutical interventions.

The second section, "Energy and Human Performance", reports the physiological assumptions of energy production in human performance.

The chapter "Methodological Procedures for Non-Linear Analyses of Physiological and Behavioural Data in Football" summarizes the main methods of extracting positional data using non-linear analyses such as entropy scales, relative phase transforms non-linear indices, cross-correlation, fractals, and clustering methods. A football match is a complex and dynamic system with emergent behavior, self-similarity, self-organization and a chaotic component. Physiological and behavioral football data should therefore be integrated to determine the complexity and non-linearity of the system.

The chapter "Justification of some Aspects of the Choice of Training Means Selection in Track-and-Field Jumps" establishes that the take-off in track-and-field jump training exercises uses relatively independent biomechanics (BM), similar to competitive jumping. A study of the underlying biomechanics and physiological processes can improve technique development choices. Four types of take-off training are identified, each with specific features: (i) involving the BM of legs (take-off leg) and body extension; (ii) involving the swinging links BM; (iii) involving the "overturned pendulum" BM; and (iv) involving the swinging links BM combined with the "overturned pendulum" BM.

"From Exercise Physiology to Network Physiology of Exercise" presents new insights into Exercise Physiology based on Molecular Exercise Physiology and Integrative Exercise Physiology. Inspired by the new field of Network Physiology and Complex Systems Science, Network Physiology of Exercise, the dynamics of the vertical and horizontal physiological network interactions were explored in new theoretical assumptions, research programs, and the practical applications of Exercise Physiology.

"Energy Cost of Walking and Running" examines walking and running as a basic way of influencing an individual's condition, health, and fitness. Speed of movement and the energy cost (EC) of applied physical activities can be measured in the laboratory and in the field, enabling movement training to be speeded up. The total EC needs to exceed the so-called stimulus threshold, that is, the subject of physical training must reach a certain minimum level of total EC of applied physical training. Assessing the energy intensity and cost can significantly contribute to reducing fatigue and the incidence of injury, and to increasing physical fitness.

The chapter "Mechanical Limits of Cardiac Output at Maximal Aerobic Exercise" discusses the factors that limit the maximum aerobic exercise by maximal oxygen uptake (VO2max). Cardiac output is the dominant determinant of VO2, so the improvement in VO2max depends largely on the increase of the return function and the effect of muscle contractions. Stroke volume is a key determinant of maximum cardiac output and therefore has extensive implications for peak performance. Significant academic attention has been paid in recent years to the physiological processes associated with altitude training, and to the role of exercise in the clinical conditions leading to hypoxia.

The third and final section, "Physical Exercise and Immune Response", explores the immune response to physical activity and exercise in a clinical perspective. Exercise training and regular physical activity have been suggested as non-pharmacological approaches to improving breast cancer outcomes. Research reports improvements in cardiorespiratory fitness levels, the completion rate of pharmacological thera-pies, reduction of cancer-related fatigue, and the improvement of muscle strength and quality of life as among the positive interdependencies between exercise train-ing and breast tumour clinical outcomes.

The chapter "Exercise Training in the Spectrum of Breast Cancer" reports the effects of exercise on the regulation of metabolic and steroid hormones, tumour-related inflammation, and the attenuation of cancer-induced muscle wasting, highlighting exercise designs that can trigger the best outcomes. According to the authors, it is difficult to identify, among diverse exercise protocols, those that produce the most effective outcomes for breast cancer patients. However, a combi-nation of moderate to vigorous aerobic and resistance exercise appears to promote the best results.

The final chapter "Physical Activity and Vaccine Response" discusses the effect of exercise on vaccine response. The study summarizes the current understanding of exercise and antibody production.

We, therefore, invite readers to discover the most recent publications, research, and theoretical frameworks in the fundamental area of Exercise Physiology.

Ricardo Ferraz, Henrique Neiva and Daniel A. Marinho
Department of Sports Sciences,
University of Beira Interior,
Covilhã, Portugal

José E. Teixeira and Pedro Forte
Departamento de Desporto e Educação Física,
Instituto Politécnico de Bragança,
Bragança, Portugal

Luís Branquinho
Higher Institute of Educational Sciences of the Douro,
Porto, Portugal

Exercise, Brain and Cognitive Functioning

Chapter 1

The Performance during the Exercise: Legitimizing the Psychophysiological Approach

Ricardo Ferraz, Pedro Forte, Luís Branquinho, José E. Teixeira, Henrique Neiva, Daniel A. Marinho and Mário C. Marques

Abstract

Over the years, there has been a growing interest in the study of issues related to the psychophysiological processes underlying sports performance. A relatively recent perspective is supported by the concept that the brain acts as a central regulator of performance during exercise. This phenomenon is called pacing and is based on the premise that prior knowledge about the activity plays a fundamental role for individuals to self-regulate their efforts throughout the exercise. However, knowledge regarding this topic remains scarce, and further clarification is needed. This chapter reports new perspectives in relation to the existing evidence regarding the role of the brain as a central regulator of performance, questioning the complex interdependencies and interrelations between fatigue and physical exercise in the light of a psychophysiological perspective. A broader understanding of the cognitive basis of the psychophysiological phenomenon during the exercise is needed, bringing together concepts such as pacing behavior, decision-making, self-regulation of effort, prior knowledge of the duration of the task, and perception of effort.

Keywords: psychophysiological, brain, fatigue, pacing, performance

1. Introduction

Exercise is characterized as a complex activity, in which the phenomenon of fatigue is enigmatic and stimulating, therefore requiring further investigation [1–4]. Over time, efforts have been made to study this phenomenon in the field of sports sciences [5–7]. However, knowledge about fatigue remains ambiguous, unpredictable, and difficult to fully explain. There is a wide range of variables (training load, anxiety, etc.) that can affect the fatigue process during exercise and its synergies with the human body responses [8, 9]. For these reasons, there is no consensus in the scientific community regarding fatigue during exercise [4, 10–12], and therefore, there is no unique definition for the concept of exercise fatigue. Thus, reaching a single definition remains a scientific challenge.

Until today, classical fatigue theories continue to be the main focus of discussion on the subject. However, recent studies have emerged, identifying flaws and limitations in these theories, essentially because they do not consider significant factors in their analysis [1, 2, 4]. Following this line of investigation, the concept of fatigue has been evidenced in new investigations as a result of other aspects [13].

So far, the fatigue concept was specially based on physiological variables [14]. The new data do not fully support innovative approaches in relation to the phenomena considered until then. In fact, the studies suggested expanding the scope and focus of the fatigue research [2]. This is because the physiological perspective justifies part of the problem [15]. However, the remaining problem seems to be explained by multifactorial variables, raising the possibility of new perspectives and psycho-physiological approaches [16–18].

Based on this perspective, effort regulation has emerged as a choice that athletes should take during exercise and that strongly influences performance [17]. This control of effort on the part of athletes has been called "pacing," and it is assumed as a valuable concept in sport, supporting the existence of a psychophysiological perspective [16]. Although coaches and athletes are aware of the importance of pacing, it has been the object of study by researchers only in recent years, and so, it continues to have little expression in literature [19]. However, investigations have shown the existence of a psychophysiological system capable of controlling physical capacities and with apparent applicability in all types of exercises, including team sports [16]. The pacing phenomenon is directly linked to exercise, and until now, there was a propensity to look at pacing as a purely psychological phenomenon. It was not considered an object of analysis in the field of sports sciences, which mainly explores physiological phenomena associated with exercise [20]. Recent findings in brain research have shown that pacing is a phenomenon with strong interconnection between the psychological and physiological dimensions [16]. In fact, stress resulting from high-intensity exercise (which leads to exhaustion) can cause an unconscious or conscious inhibition of the athlete's tolerance to pain [21]. This may cause the central nervous system (CNS) to regulate the exercise pace as much as necessary for the athlete's pain to become bearable, allowing the task to be completed [4, 10]. Generally, researchers agree that the perceived discomfort of fatigue occurs just before the occurrence of physiological limitations in the muscles. However, the precise role of the CNS in detecting, causing, or even canceling the perception of fatigue remains unclear, and there is a gap in the literature regarding this phenomenon [12, 18]. This perspective does not come from the physiological system, it only emphasizes that effort regulation is consciously or unconsciously commanded by the brain. Thus, greater knowledge regarding the exercise operating mechanisms may bring about new approaches to explore the phenomenon of pacing in exercise [4, 10, 12, 16, 18]. Furthermore, if the pacing phenomenon acts as a regulatory system that allows the effort to be completed in the context of training and competition, factors such as previous experience and perfect knowledge of the task appear to be fundamental to success [4, 10, 16]. In fact, prior knowledge about the activity to be performed (i.e., duration, importance, demands of the game) is assumed to be a fundamental point to support a psychophysiological approach to the phenomenon of pacing [9, 18, 22]. That is how athletes are able to self-regulate their own performance during exercise [19].

Athletes' previous knowledge about the duration of the task can induce changes in the performance, as demonstrated in ultra-marathon running [23]. In addition, the circumstance in which fatigue is expressed is another interesting factor for investigation, although little explored. The sports modalities' specificities are important and help to interpret the results and sensations of perceived fatigue [16, 18]. This is because the type of required effort differs between modalities (i.e., collective or individual) [19]. Upon that, different reflections about the mechanisms of fatigue are necessary. It is also important to note that there is a research gap considering this subject in team sports, since the existing studies on the phenomenon of pacing are mostly related to endurance sports [16, 18, 19]. This may be related to the intrinsic and specific characteristics of each team sport. It can create issues in the analysis, possibly provoking a propensity for the study of specific variables, which could be considered as a reductionist approach to the problem. That said, based on the lack of identified research, it becomes important to enhance knowledge about the mechanisms of fatigue in sport.

2. The fatigue in sport and exercise

As presented above, there are several concepts about fatigue; nevertheless, it can be defined as "extreme tiredness after effort; reduction in the efficiency of a muscle or organ after prolonged activity" [24]. Previous research has shown that the fatigue phenomenon is more complex and has a broader spectrum of parameters, and more comprehensive concepts are necessary. The wide range of investigations carried out had a beneficial impact in expanding the available definitions of fatigue. It is difficult to accurately determine the development of fatigue as a concept in the sports sciences [1]. At the beginning of the twentieth century, two phenomena were identified that helped to characterize fatigue; (i) the reduction of muscle strength; (ii) tiredness as a sensation [25]. In fact, over time, there have been variations in the definitions of fatigue in the field of sports sciences such as; (i) "the difficulty in maintaining the necessary or expected strength" [26]; (ii) "the decreased ability to generate maximum strength" [27]; (iii) "a reversible state of strength depression, including a lower rate of strength increase and slower relaxation" [28]. That said, it is difficult to gather consensus on a single definition of fatigue. This is problematic when trying to consistently compare and interpret the different concepts. However, these inconsistencies allow the debate on the topic to be constantly open, mainly on its usefulness and applicability in the different modalities. Moreover, reaching an early definition of fatigue, which is only accepted for lack of alternatives, may help to reiterate the complexity of the fatigue phenomenon [1, 2].

2.1 Peripheral fatigue and central fatigue

Fatigue in sport and exercise can theoretically be categorized into two types (i.e., peripheral fatigue and central fatigue) [29, 30]. Peripheral fatigue is related to decreased muscle strength production caused by processes distal to the myoneural junction [31]. The concept of peripheral fatigue originated from studies carried out in the 1920s [32–34]. These studies led to the conclusion that, just before the end of the exercise, the muscles' requirement for oxygen exceeded the heart's ability to supply that oxygen. This process develops an anaerobiosis in the muscles in activity, causing the accumulation of carboxylic acid. Due to this change in the intramuscular environment, the continuation of the contraction becomes impossible, and therefore, the muscles reach a state of failure. These pioneering studies claim that fatigue is the result of increased intramuscular carboxylic acid, which is produced within the body only under anaerobic conditions [1]. These conclusions were supported by the fact that exercise performance improves with oxygen inhalation [35]. Furthermore, the authors also concluded that the main limiting factor in exercise tolerance was the heart's ability to pump blood to active muscles. Thus, fatigue is possibly a consequence of the heart's inability to supply oxygen and the cardiovascular system not being ready to remove waste through the oxidation of active muscles [1]. Based on this perspective, the cardiovascular system appears to restrict performance due to the difficulties induced by the breakdown of the supply of blood, nutrients, and oxygen to the active muscles [36, 37]. Insufficiencies in the heart's pumping capacity, as well as the reduced density of capillaries, can limit the amount of ventilated blood that reaches the muscles, consequently limiting performance. This theory, known as the anaerobic/cardiovascular/catastrophic model of human performance exercise, predicts the failure of cardiac homeostatic balance [32–34]. Although this model has been criticized, this theory has prevailed and is probably still the most cited theory of exercise-induced fatigue, but some limitations are known and have been the subject of previous analysis.

An investigation [38] concluded that the maximum cardiac output limit of the heart was reached via the evolution of myocardial ischemia, when the heart loses the ability to pump more blood because it has reached the rate of oxygen consumption limit. Additionally, this investigation showed that obtaining a higher flow rate limits the blood flow to the muscles in activity, inflicting the anaerobiosis and limiting the capacity to remove lactic acid. The increase in the concentration of lactic acid directly influences the contractile capacity of muscle fibers, presenting an association with mechanisms that induce muscle fatigue. In addition, the same authors also considered the prospect that myocardial ischemia, as a result of reaching maximum cardiac output, which is a limiting factor in exercise and a threat to the integrity of cardiac tissue, can be avoided due to the existence of a governor, located in the brain or the heart, which protects from possible damages.

Even so, there is still a lack of scientific evidence showing that muscles' energetic profile actually becomes anaerobic during exercise and close to fatigue; or even that oxygen consumption or cardiac output consistently reaches a peak. This peak would be a requirement for its implication in fatigue during maximal intensity exercises [1]. In addition, a healthy heart, even during maximal intensity exercises, does not assume the existence of myocardial ischemia. Upon that, the hypothesis about an existing regulator in the brain or in the heart lacks scientific support. The model also suggests that peripheral fatigue events would lead the brain to recruit additional muscle fibers in an effort to help these fatigued fibers. Thus, to maintain the intensity of the exercise, it will be necessary to engage more and more available muscle fibers at their maximum capacity. However, this prediction is contradictory to other aspects of the model, as the continuous muscular recruitment should aggravate the metabolic crisis that the model foresees to be the reason for the end of the exercise [29, 30]. The main issue of this theory is that fatigue is a catastrophic event, sustained by an assertive response, leading to the total failure of the active muscles to continue to produce strength. However, catastrophic muscle or organ failure clearly does not occur in exhaustion for healthy individuals during any type of exercise [1]. Additionally, for this model, fatigue is shown from the perspective of an exhaust failure system. However, skeletal muscle fibers are never fully recruited during exercise; muscle adenosine triphosphate (ATP) never falls below 60% of resting levels, and glycogen concentration decreases but is not depleted during exercise [39]. Even more, in many circumstances, fatigue occurs before high concentrations of metabolites, such as lactate, H^+, extracellular K^+, without disturbances in muscle Ca^{2+} kinetics and without high core temperatures or significant hypohydration [40].

All these observations contradict the prediction of the peripheral linear catastrophic model, which states that some type of homeostatic failure should occur to cause fatigue. However, the importance of the model peripheral component remained. In the Hill's model (presented in the beginning of the twentieth century) [32, 34], the physiological aspect is accepted; although it proves unable to respond to the complexity of the fatigue phenomenon in a broader scope. The model considers a refutable role of the brain to disrupt myocardial ischemia; however, it ignores the role of neural control over all physiological systems [1]. While peripheral fatigue occurs through processes outside the CNS, it is believed that the origin of central fatigue lies in the CNS, with the loss of muscle strength occurring through processes proximal to the myoneural junction. Specifically, this refers to sites within the brain, spinal nerves, and motor neurons, and it is related to instances in which the CNS presents a decrease in neural impulse to the muscle [41]. Central fatigue is perceived as the failure of the central nervous system to drive the muscle to its maximum, resulting in some loss of strength [42]. The decrease in strength/performance production [43] is largely justified within the central nervous system (brain and spine—central) and anywhere outside the central nervous system (e.g., peripheral muscle). Comparatively, little research has been done on the role of the CNS in fatigue until the last decades [44], which is curious considering that it has long been suspected of being a central component of fatigue [1]. The impact of the research findings on peripheral fatigue and the limitations to measure central fatigue due to the lack of objective and direct tools explain the current research gap in this field. In fact, central fatigue is usually only accepted when experimental findings do not support any peripheral cause of fatigue [44].

The central nervous system plays an important role in maintaining homeostasis [45]. Therefore, the motor component of the brain is responsible for the production of motor drive and the recruitment of motor units during exercise [45]. Thus, the brain takes control of cognition and recognition of physical sensations that are perceived as fatigue. Perceived fatigue results from exercise, and it is felt as a "sensation" (common/frequent) during exercise. The workload can create a sensation so intense that it is perceived as a need to reduce the strength to successfully complete the activity (i.e., pacing). In some cases, it may be necessary to stop exercising altogether if the sensations felt are too intense [46]. For this reason, the various stages that athletes go through during exercise are indicative that physiological mechanisms are not the only ones responsible for regulating exercise intensity. Also, humans exhibit an anticipated aspect of exercise regulation, possibly with regard to factors such as perception of the effort required for the task, and motivation [1]. Physical and biochemical changes during exercise are physiological aspects that naturally must be considered. However, perceived fatigue should also be carefully considered due to the influence in behavior/performance. Therefore, it is important to study perceived fatigue with similar importance [1, 45, 46]. Considering that the catastrophic failure of the system does not occur, there is a possibility for the appearance of a psycho-physiological model [16].

2.2 Psychophysiological evidence

The inability of the peripheral and central fatigue processes to convincingly explain sport and exercise fatigue allowed the researcher to predict explanations for the fatigue phenomenon [29, 30]. An interesting perspective that has recently emerged is the concept of the brain acting as a central regulator of the exercise performance [4].

As mentioned above, the peripheral catastrophic model remains the dominant model, and this is essentially due to the modifications made to the Hill's model [32–34] with the incorporation of factors such as energy supply and depletion [47]. In fact, the introduction of energy supply to the model suggested that high-intensity exercise was due to the inability to provide ATP at rates fast enough to maintain the exercise high intensities [47]. Based on this model, the training process and the diet generate an increase in the storage capacity (for example, glycogen), and the increased use of metabolic substrates during exercise may result in a higher production capacity of ATP. Controversially, in the variation of energy depletion of the model, it was suggested that the amount of carbohydrates was the limiting factor [47]. This is probably due to the finding that fatigue during prolonged exercise is strongly associated with significant reductions in the liver and muscle glycogen [37, 48]. In addition, improvements in tolerance to hypoglycemia as a result of exercise allow exercise to continue [48]. Nevertheless, none of the models are fully accepted.

The concept of the existence of a central regulator, responsible for regulating muscle metabolic activity and performance via peripheral afferent feedback, was reintroduced by Ulmer in 1994 [49]. The author suggested that this central regulator anticipates the end point of an exercise. Anticipation was based on previous experience of the same exercise or knowledge of the duration of the task, which regulates the metabolic demand from the beginning of the exercise. This allows the task to be completed without catastrophic physiological failure. The control of the metabolic demand regulated by the brain is called teleanticipation. This central governor evidenced by Ulmer [49] has been analyzed in several research studies [40, 41, 45, 50, 51]. These studies were the starting point for the appearance of the so-called anticipatory feedback model of exercise regulation [4], which constitutes a psychophysiological approach of the fatigue phenomenon. This model assumes that the exercise is self-regulated from the beginning by the athletes based on previous experiences, knowledge of the expected distance, duration of the current exercise, and afferent physiological feedback regarding some variables (i.e., muscle glycogen levels, skin and body temperature) [40, 41]. Processing this information allows the brain to predict and regulate the most appropriate exercise intensity allowing an optimal performance without serious homeostatic disruptions [51]. These predictions are similar to the model that classified the perceived effort (RPE). Moreover, the physical, mechanical, and biomechanical variables required during exercise are constantly monitored by the brain, and it is through this afferent feedback that the athlete's conscious RPE arises. During exercise, conscious RPE is continuously compared with standard RPE and will progressively increase and reach its desired maximum at the expected end of the exercise. The intensity of the exercise is modulated according to an acceptable level that the brain interprets as tolerated, taking into account the continuous comparison between the standard RPE and the real, conscious RPE [50, 52].

The anticipatory feedback model defends that fatigue, instead of a physical state, is a conscious sensation generated from the interpretation of subconscious regulatory processes [45, 51]. It is also suggested that RPE is not simply a direct manifestation of afferent physiological feedback, but that it also plays an important role in preventing excessive intensity of exercise duration. It acts as the motivating element behind the athlete's decision to completely stop the exercise or adjust the intensity to guarantee its completion without significant or harmful physical damage [4]. Despite the lack of experimental research on the subject, some phenomena support this fatigue model [53].

2.3 The concept of pacing

Recently, numerous studies investigated the interaction between cognition and sports performance [54, 55]. The pacing behavior has been widely identified as an essential component of success in many sports and is directly related to a high spectrum of cognitive skills [56–61]. Pacing has been described as a multifaceted process that requires a set of decision-making in which athletes need to decide when and how they will distribute their available energy throughout an exercise [60, 62, 63].

The ideal pacing behavior in time trial competitions is characterized by: the balance between the quick start to make optimal use of energy resources; preventing negative changes in performance resulting from early fatigue; and inefficient energy losses associated with speed fluctuations during the race [64]. To determine the most appropriate pacing behavior, a set of variables (i.e., biomechanical, physiological, psychological, and environmental) [46] are crucial to maintain internal homeostasis [65] and to avoid premature burnout [66, 67].

The concept of pacing supports the anticipatory feedback model and cannot be investigated from a purely physiological perspective [62]. The effort distribution is part of the exercise, which suggests that voluntary behavior (effort) may limit performance rather than the absolute capacity of a single physiological system [46]. The role of a central process and how it will be executed must be considered when developing a pacing behavior.

A heuristic model of decision-making was developed to integrate the theories of decision-making and pacing, in which heuristics were considered intuitions that require low cognitive demands [20]. However, the heuristic decision-making model did not consider the connection between perception and action that takes place in tactical pacing environments, in which some actions depend on the opponents' behavior [20, 63, 68]. Later, a detailed explanation of the pacing phenomenon emerged and was presented as a behavioral result of the decision-making process and included human-environment interactions. Pacing leads athletes to make decisions in complex and demanding environments, where they are successively encouraged to modify, choose, and evaluate their behavior [69]. The brain and cognitive processes interact and act as an information processing system [70].

The competitive environments such as the stage of competition [71], the importance of competition, and the probability of qualifying time [72] can modify the athlete's pacing behavior. During the competition, opponents are the most common affordances. However, there are other environmental factors that can influence the pacing behavior of athletes. Factors such as music [73], performance feedback [74], and weather conditions [75] can lead to voluntary reductions in exercise intensity. These reductions in intensity occur before any real physical need to do so and before the performance compromise occurs as a result of any failure of the physical system [76, 77].

The presence of pacing behavior in sports is important regarding the view of the anticipatory regulation proposed for the performance of the exercise. It seems that athletes perform the exercise less effectively when performing an exercise that is unfamiliar and whose demands are not entirely clear [78, 79].

Changes in exercise intensity during resistance exercises were reported in the initial phase of exercise before any peripheral physiological cause of fatigue [80]. These data suggest that the modification of exercise intensity (pacing) during exercise occurs in anticipation and not as a result of stress or failure of the physiological system [62]. Thus, the pacing strategies may be used to guarantee the completion of the exercise without any physical damage. The previous experience and knowledge of the demands of the exercise will play an important role [62].

The use of pacing strategies during exercise can provide support for components of the anticipatory feedback model [81, 82] as well as refute aspects of the peripheral linear catastrophic model. During the self-regulated exercise, it is observed that the pacing behavior depends on the environment, the demands, and objectives of the exercise and the afferent physiological feedback [83]. This is in agreement with the anticipatory feedback model. If an athlete's pacing behavior is determined by the accumulation of metabolic products or depletion of energy reserves, as predicted by the peripheral linear catastrophic model, athletes would always begin exercising at an unsustainable pace [49, 74, 84–86]. Gradually, they would slow down due to the negative effect of peripheral variables, which is not actually put into practice. The peripheral linear catastrophic model states that the only possible stimulation behavior in exercise is linear [27, 29, 30, 45, 67]. The model simply does not allow the existence of other strategies. However, the evidence for these other strategies is abundant [16, 19].

Previous exercise knowledge/experience can be important information that the brain uses to select a more appropriate exercise intensity. Research on the use of pacing strategies in exercises has confirmed that the precision ability of pacing is improved with training and experience [84, 87].

2.4 The end spurt phenomenon

The end spurt phenomenon supports the anticipatory feedback model [23] and is characterized by a substantial increase in the intensity of the exercise when it approaches the end. This model disregards the effort during the entire exercise. Throughout the exercise, there is often a level of uncertainty about the precise end point of the exercise period and the type of effort that will need to be spent until the end. These aspects are responsible for influencing the athlete's pace and can, at any time, force the athlete to make changes in the pace, which cannot be predicted before exercise. This type of uncertainty can result in the maintenance of a motor unit and metabolic reserve throughout the exercise [4, 82, 88]. The athlete cannot be sure of what can happen in the rest of the exercise period and (unconsciously) retains some type of reserve to remain prepared to respond to any potential physical challenges. This may allow the accomplishment of the exercise without significant interruption of homeostasis. When the end of the exercise approaches, the uncertainty decreases and the accumulated reserve is no longer accurate, so that the athlete can significantly increase the metabolic demand increasing speed/power. Actually, this is a possible evidence that fatigue is not caused by the inability of muscles to produce strength [23].

2.5 The knowledge about the exercise duration

Knowledge of exercise duration as a regulator of exercise performance plays an important role and is supported by investigations about the knowledge or no knowledge of the task duration [1, 79, 89–91]. This type of research is usually called "deception," in which participants believe that the exercise will last for a certain length of time; however, at the end of that period, they are asked to continue exercising. In one of the first researches on the topic [90], participants were asked to run on a treadmill at 75% of their maximum speed. However, in the first phase, they were asked to run for 20 min and were interrupted after 20 min. In a second phase, they were asked to run for 10 min, and at the end of this period, they were asked to run for 10 min more. In a third phase, they were asked to run, but were not told for how long (they were stopped after 20 min). All phases were performed at the same running speed and lasted 20 min. The results indicated that the participants' RPE had increased significantly between 10 and 11 min in the 10 min deception. The deception occurred immediately after revealing the information that the participants were required to continue the exercise for a longer time. These changes in the perception of effort and pleasure occurred despite the fact that there were no changes in running speed or in physiological responses to the exercise period. The significant increase in RPE after the participants were asked to prolong the exercise was also found in another research with similar protocols [91], reporting that the effect also increased in the last minutes of exercise, probably because the participants were aware that the exercise was close to ending. These findings are related to the end spurt phenomenon. An increase in feelings of pleasure at the end of the exercise may explain the end spurt happening. Furthermore, there was no increase in the effect on the trial when the participants did not know the duration of the exercise, and the effect continued to decrease throughout the trial [91].

A recent study [79] assessed how the manipulation of knowledge about the duration of a training task restricts the pace and tactical behavior of soccer players during the performance of small-sided games (SSG). Players were instructed to play the SSG for 10 min, but after completing the 10-min game, they were asked to play for another 10 min, and in another situation, they were previously informed that they would play for 20 min. The results indicate that the first 10 min of each scenario had a greater physical impact regardless of the initial information that had been revealed. During that time, tactical behavior has also showed greater variability. In addition, there was an increase in distance from teammates during the second 10-min period in which the duration was fully known. That may be due to a smaller pacing behavior. This study showed that prior knowledge of the duration of the task led to different physical and tactical behaviors of the players, and these data have been corroborated by other investigations [92–94]. These findings confirm the possibility of changes in the pacing patterns of the players, as a consequence of the knowledge of the duration of the task that leads to consider the possibility of the nonlinearity of the fatigue effect previously reported in other studies [23, 95]. These data suggested that the knowledge of the exercise duration assumes a fundamental role for the adequate regulation of the exercise performance, as in the anticipatory feedback model. The increases seen in athletes' RPE when deception is revealed may reflect an interruption of the feed-forward/feedback mechanism, which is fundamental to RPE as suggested by other studies [4, 96]. Moreover, it is also verified that both the RPE and the physiological responses (oxygen consumption, heart rate) present lower values when the duration of the exercise is not known compared with the moments when the duration is known. However, no significant differences in exercise intensity were found [90, 91]. Thus, these responses may reflect a subconscious improvement in the effort economy in order to retain energy due to the unknown duration of the exercise period. That said, knowledge of the end point of the exercise plays a great role in the perceptual and physiological responses to that same period of exercise [97]. This fact is further evidenced by the observation that the responses of the RPE to the exercise are robust when the duration of the exercise is known, even when no information is provided to the athlete about the exercise [85].

Research results related to exercise duration prior knowledge provide additional evidence on some processes by which athletes can retain physiological reserves during exercises of uncertain duration [91]. These findings provide support for a central role of the CNS in regulating exercise performance [98], probably to ensure the maintenance of homeostasis and the guarantee of an emergency "reserve" of energy/physical capacity [12, 91].

2.6 The relationship between RPE and performance

The perceived effort during the exercise is reflected in changes in the sensation to regulate the athlete's physical integrity [99]. The output (perceived effort) is based on a combination of sensory inputs and cognitive processes [100]. One of the most accepted parameters during the exercise is the RPE response, which represents a sum of afferent feedback signals [100] and supra-spinal mechanisms [86]. Based on the principles of self-regulation, it is suggested that the use of RPE methods to monitor training presents itself as an effective tool for all types of exercise.

RPE has been recognized as a valid and reliable indicator of the level of physical effort by the American College of Sports Medicine. The RPE characterizes the conscious perception of the effort experienced during the exercise, which gives it a considerable practical value for the athlete. Thus, exercises that require higher levels of energy expenditure and physiological effort usually present higher RPE. Furthermore, previous studies [4, 20, 63] reported the existence of a relationship between variations in the RPE during the exercise and the duration of the exercise, which highlights the assumptions of the anticipatory feedback model. They also suggest that the RPE is effectively a crucial regulator of the exercise performance. Additionally, the suggestion that RPE may vary from the beginning of the exercise through changes in the ambient temperature and intensity of the exercise [4, 101] before effective physiological changes supports the role of the RPE in anticipatory regulation of the exercise. This evidence suggests that RPE may not be a direct reflection of the athlete's physiological state during exercise, but rather an anticipatory sensory regulator of exercise performance. Upon that, the RPE may undergo variations during the exercise in anticipation of the occurrence of physiological changes and not because of these changes. This supports the fact that the nature of fatigue should be considered as previously reported [90].

2.7 Other approaches

An interesting alternative to the brain regulation model has been proposed [11]. In this approach, there is an acceptance that the brain regulates muscle recruitment and limits performance; however, there is also reticence about the need for a central regulatory governor. This perspective suggests that the search for a central governor in the brain's subconscious may be similar to current reductionist approaches that search for a single cause of fatigue. The anticipatory feedback model states that a central regulator in the brain maintains subconscious control over skeletal muscle fibers recruitment during exercise. However, the presence of a single region of the brain, exclusively dedicated to regulating exercise performance, is highly unlikely. It is antagonistic regarding everything known about the functioning of the brain as an integrated organ of maximum complexity where each region contributes to the general functioning of the brain [62]. This may also explain why the specific region of the brain considered the central governor was not found. This model also states that the perception of effort is fundamental to demote the individual to continue in dangerous levels of conscious exercise that can be theoretically redundant. This is because the subconscious regulator will prevent athletes from exercising at a dangerous level regardless of the motivation that may exist to continue. However, another author [4] states that the anticipatory feedback model could exist without the perception of effort being considered. An alternative suggestion was given with a simplified model that helps to explain some of the evidence attributed to the anticipatory feedback model. This model determines that the end of the exercise occurs when the effort required to continue exercising is similar to the maximum effort that the individual is willing to provide or when the individual believes that he/she has provided a true maximum. Therefore, the subject realizes that it is not viable to continue exercising [11].

The increase in the effort that the individual is willing to put into the exercise will improve his tolerance, as long as it does not exceed what the individual understands as his maximum effort [11]. The importance of the perception of effort remains clear, but the existence of a central regulator in the brain is not necessary. Additionally, it has been suggested that the gradual increase in RPE over time and at different rates in response to changes in exercise intensity and ambient temperature can be explained by other factors; a central regulator that uses perceived exertion as a mechanism of security would be an insufficient explanation. The RPE is generated through signals originating in the CNS, specifically referring to prolonged submaximal exercises with a constant workload [31]. These data are highlighted and refuted in other investigations [10, 11]. It was demonstrated that the RPE suffered changes almost since the beginning of the exercise as a result of the verified differences in the exercise intensity and in the ambient temperature. In addition, it is important to note that the increase in the CNS motor commands could not happen without afferent sensory feedback, which is similar to the anticipatory feedback model [10].

3. Conclusion

This chapter provided a broader understanding of the cognitive basis of the psychophysiological phenomenon during the exercise, bringing together concepts such as pacing behavior, decision-making, self-regulation of effort, prior knowledge of the duration of the task, and perception of effort. This reinforced the role and contribution of the cognitive component in the pacing behavior. Furthermore, the development of fatigue during exercise seems to result from a complex interaction between the physical and psychophysiological components responsible for changes in the exercise intensity, and it can be pointed out that central and peripheral fatigue can help to exercise the intensity regulation at the beginning of a rhythmic self-paced. Also, the perceived responses may be of higher importance to control the intensity of the exercise, especially in the final phase, this results from the attempt to retain a reserve of energy that allows maximum effort in the end. About the prior knowledge about the task duration, it can create a greater capacity to regulate the effort, leading athlete to better manage energy reserves throughout the exercise. It would be also interesting to continue to analyze the impact of psychophysiological factors on the perception and regulation of fatigue by team sports players according to recent studies of psychophysiological fatigue.

Acknowledgements

This work is supported by national funding through the Portuguese Foundation for Science and Technology, I.P., under the project UID04045/2020.

Conflict of interest

The authors declare no conflict of interest.

Author details

Ricardo Ferraz[1,2*], Pedro Forte[2,3,4], Luís Branquinho[3], José E. Teixeira[2,5], Henrique Neiva[1,2], Daniel A. Marinho[1,2] and Mário C. Marques[1,2]

1 University of Beira Interior, Covilhã, Portugal

2 Research Center in Sports Sciences Health Sciences and Human Development, Covilhã, Portugal

3 Douro Higher Institute of Educational Sciences, Penafiel, Portugal

4 Instituto Politécnico de Bragança, Bragança, Portugal

5 University of Trás-os-Montes e Alto Douro, Vila Real, Portugal

*Address all correspondence to: ricardompferraz@gmail.com

IntechOpen

References

[1] Noakes TD. Fatigue is a brain-derived emotion that regulates the exercise behavior to ensure the protection of whole body homeostasis. Frontiers in Physiology. 2012;3:82. DOI: 10.3389/fphys.2012.00082

[2] Marino FE, Gard M, Drinkwater EJ. The limits to exercise performance and the future of fatigue research. British Journal of Sports Medicine. 2011;45:65-67. DOI: 10.1136/bjsm.2009.067611

[3] Smith MR, Marcora SM, Coutts AJ. Mental fatigue impairs intermittent running performance. Medicine and Science in Sports and Exercise. 2015;47:1682-1690. DOI: 10.1249/MSS.0000000000000592

[4] Tucker R. The anticipatory regulation of performance: The physiological basis for pacing strategies and the development of a perception-based model for exercise performance. British Journal of Sports Medicine. 2009;43:392-400. DOI: 10.1136/bjsm.2008.050799

[5] Halson SL. Monitoring training load to understand fatigue in athletes. Sports Medicine. 2014;44:139-147. DOI: 10.1007/s40279-014-0253-z

[6] Borresen J, Ian LM. The quantification of training load, the training response and the effect on performance. Sports Medicine. 2009;39:779-795. DOI: 10.2165/11317780-000000000-00000

[7] Russell S, Jenkins D, Smith M, Halson S, Kelly V. The application of mental fatigue research to elite team sport performance: New perspectives. Journal of Science and Medicine in Sport. 2019;22:723-728. DOI: 10.1016/j.jsams.2018.12.008

[8] Jones CM, Griffiths PC, Mellalieu SD. Training load and fatigue marker associations with injury and illness: A systematic review of longitudinal studies. Sports Medicine. 2017;47:943-974

[9] Balagué N, Hristovski R, Almarcha MDC, Garcia-Retortillo S, Ivanov PC. Network physiology of exercise: Vision and perspectives. Frontiers in Physiology. 2020;11:1607

[10] David Noakes T, Tucker R. Do we really need a central governor to explain brain regulation of exercise performance? A response to the letter of Dr. Marcora. European Journal of Applied Physiology. 2008;104:933-935. DOI: 10.1007/s00421-008-0842-3

[11] Marcora SM. Do we really need a central governor to explain brain regulation of exercise performance? European Journal of Applied Physiology. 2008;104:925-929. DOI: 10.1007/s00421-008-0818-3

[12] Swart J, Lindsay TR, Lambert MI, Brown JC, Noakes TD, Robert Lindsay T, et al. Perceptual cues in the regulation of exercise performance—physical sensations of exercise and awareness of effort interact as separate cues. British Journal of Sports Medicine. 2012;46:42-48. DOI: 10.1136/bjsports-2011-090337

[13] Kalkhoven JT, Watsford ML, Coutts AJ, Edwards WB, Impellizzeri FM. Training load and injury: Causal pathways and future directions. Sports Medicine. 2021;51:1137-1150

[14] Djaoui L, Haddad M, Chamari K, Dellal A. Monitoring training load and fatigue in soccer players with physiological markers. Physiology & Behavior. 2017;181:86-94

[15] Mujika I, Halson S, Burke LM, Balagué G, Farrow D. An integrated, multifactorial approach to periodization for optimal performance in individual and team sports. International Journal

of Sports Physiology and Performance. 2018;**13**:538-561

[16] Edwards A, Polman R. Pacing in Sport and Exercise: A Psychophysiological Perspective. New York: Nova Science Publishers; 2012. Available from: https://researchonline.jcu.edu.au/22922/ [Accessed: January 10, 2021]

[17] Gabbett TJ, Walker B, Walker S. Influence of prior knowledge of exercise duration on pacing strategies during game-based activities. International Journal of Sports Physiology and Performance. 2015;**10**:298-304. DOI: 10.1123/ijspp.2013-0543

[18] Waldron M, Highton J. Fatigue and pacing in high-intensity intermittent team sport: An update. Sports Medicine. 2014;**44**:1645-1658. DOI: 10.1007/s40279-014-0230-6

[19] Thompson K. Pacing: Individual Strategies for Optimal Performance. Champaign: Human Kinetics; 2014

[20] Renfree A, Martin L, Micklewright D, St Clair Gibson A. Application of decision-making theory to the regulation of muscular work rate during self-paced competitive endurance activity. Sports Medicine (Auckland, NZ). 2014;**44**:147-158. DOI: 10.1007/s40279-013-0107-0

[21] Marcora SM, Staiano W. The limit to exercise tolerance in humans: Mind over muscle? European Journal of Applied Physiology. 2010;**109**:763-770

[22] Ferraz R, Gonçalves B, Van Den Tillaar R, Jiménez Sáiz S, Sampaio J, Marques MC. Effects of knowing the task duration on players' pacing patterns during soccer small-sided games. Journal of Sports Sciences. 2018;**36**:116-122. DOI: 10.1080/24733938.2017.1283433

[23] Millet GY. Can neuromuscular fatigue explain running strategies and performance in ultra-marathons?: The flush model. Sports Medicine (Auckland, NZ). 2011;**41**:489-506. DOI: 10.2165/11588760-000000000-00000

[24] Online OE. Help|Oxford English Dictionary 2010. Available from: https://public.oed.com/help/ [Accessed: January 10, 2021]

[25] Mosso A. Fatigue. 2nd ed. London/New York: Swan Sonnenschein & Co. Ltd; 1906

[26] Edwards RH. Human muscle function and fatigue. Ciba Foundation Symposium. 1981;**82**:1-18. DOI: 10.1002/9780470715420.ch1

[27] Bigland-Ritchie B, Furbush F, Woods JJ. Fatigue of intermittent submaximal voluntary contractions: Central and peripheral factors. Journal of Applied Physiology. 1986;**61**:421-429. DOI: 10.1152/jappl.1986.61.2.421

[28] Fitts RH, Holloszy JO. Effects of fatigue and recovery on contractile properties of frog muscle. Journal of Applied Physiology Respiratory Environmental and Exercise Physiology. 1978;**45**:899-902. DOI: 10.1152/jappl.1978.45.6.899

[29] Carroll TJ, Taylor JL, Gandevia SC. Recovery of central and peripheral neuromuscular fatigue after exercise. Journal of Applied Physiology. 2017;**122**:1068-1076

[30] Zając A, Chalimoniuk M, Maszczyk A, Gołaś A, Lngfort J. Central and peripheral fatigue during resistance exercise–A critical review. Journal of Human Kinetics. 2015;**49**:159

[31] Ament W, Verkerke G. Exercise and fatigue. Sports Medicine. 2009;**39**:389-422. DOI: 10.2165/00007256-200939050-00005

[32] Hill AV, Long CNH, Lupton H. Muscular exercise, lactic acid and the

supply and utilisation of oxygen—Parts VII–VIII. Proceedings of the Royal Society of London Series B, Containing Papers of a Biological Character. 1924;**97**:155-176. DOI: 10.1098/rspb.1924.0048

[33] Hill AV, Lupton H. Muscular exercise, lactic acid, and the supply and utilization of oxygen. QJM. 1923;**os-16**:135-171. DOI: 10.1093/qjmed/os-16.62.135

[34] Hill AV, Long CNH, Lupton H. Muscular exercise, lactic acid, and the supply and utilisation of oxygen.—Parts IV-VI. Proceedings of the Royal Society of London Series B, Containing Papers of a Biological Character. 1924;**97**:84-138

[35] Hill L, Flack M. The influence of oxygen inhalations on muscular work. The Journal of Physiology. 1910;**40**:347-372. DOI: 10.1113/jphysiol.1910.sp001374

[36] Bassett DR, Howley ET. Limiting factors for maximum oxygen uptake and determinants of endurance performance. Medicine and Science in Sports and Exercise. 2000;**32**:70-84. DOI: 10.1097/00005768-200001000-00012

[37] Fitts RH. Cellular mechanisms of muscle fatigue. Physiological Reviews. 1994;**74**:49-94. DOI: 10.1152/physrev.1994.74.1.49

[38] Raskoff WJ, Goldman S, Cohn K. The "Athletic Heart": Prevalence and physiological significance of left ventricular enlargement in distance runners. JAMA: The Journal of the American Medical Association. 1976;**236**:158-162. DOI: 10.1001/jama.236.2.158

[39] Rauch HGL, St. Clair Gibson A, Lambert EV, Noakes TD. A signalling role for muscle glycogen in the regulation of pace during prolonged exercise. British Journal of Sports

Medicine. 2005;**39**:34-38. DOI: 10.1136/bjsm.2003.010645

[40] Noakes TD. Physiological models to understand exercise fatigue and the adaptations that predict or enhance athletic performance. Scandinavian Journal of Medicine and Science in Sports. 2000;**10**:123-145. DOI: 10.1034/j.1600-0838.2000.010003123.x

[41] Noakes TD, St Clair Gibson A, Lambert EV. From catastrophe to complexity: A novel model of integrative central neural regulation of effort and fatigue during exercise in humans: Summary and conclusions. British Journal of Sports Medicine. 2005;**39**:120-124. DOI: 10.1136/bjsm.2003.010330

[42] Taylor JL, Todd G, Gandevia SC. Evidence for a supraspinal contribution to human muscle fatigue. Clinical and Experimental Pharmacology & Physiology. 2006;**33**:400-405. DOI: 10.1111/j.1440-1681.2006.04363.x

[43] Merton PA. Voluntary strength and fatigue. The Journal of Physiology. 1954;**123**:553-564. DOI: 10.1113/jphysiol.1954.sp005070

[44] Davis JM, Bailey SP. Possible mechanisms of central nervous system fatigue during exercise. Medicine and Science in Sports and Exercise. 1997;**29**:45-57. DOI: 10.1097/00005768-199701000-00008

[45] Lambert EV, St Clair Gibson A, Noakes TD. Complex systems model of fatigue: Integrative homoeostatic control of peripheral physiological systems during exercise in humans. British Journal of Sports Medicine. 2005;**39**:52-62. DOI: 10.1136/bjsm.2003.011247

[46] Noakes TD. Time to move beyond a brainless exercise physiology: The evidence for complex regulation of human exercise performance. Applied

Physiology, Nutrition and Metabolism. 2011;36:23-35. DOI: 10.1139/H10-082

[47] Costill D, Thomason H, Roberts E. Fractional utilization of the aerobic capacity during distance running. Medicine and Science in Sports and Exercise. 1973;5:248-252. DOI: 10.1249/00005768-197300540-00007

[48] Coggan AR, Coyle EF. Reversal of fatigue during prolonged exercise by carbohydrate infusion or ingestion. Journal of Applied Physiology. 1987;63:2388-2395. DOI: 10.1152/jappl.1987.63.6.2388

[49] Ulmer HV. Concept of an extracellular regulation of muscular metabolic rate during heavy exercise in humans by psychophysiological feedback. Experientia. 1996;52:416-420. DOI: 10.1007/BF01919309

[50] St Clair Gibson A, Noakes TD. Evidence for complex system integration and dynamic neural regulation of skeletal muscle recruitment during exercise in humans. British Journal of Sports Medicine. 2004;38:797-806. DOI: 10.1136/bjsm.2003.009852

[51] Crewe H, Tucker R, Noakes TD. The rate of increase in rating of perceived exertion predicts the duration of exercise to fatigue at a fixed power output in different environmental conditions. European Journal of Applied Physiology. 2008;103:569-577. DOI: 10.1007/s00421-008-0741-7

[52] Noakes TD, Peltonen JE, Rusko HK. Evidence that a central governor regulates exercise performance during acute hypoxia and hyperoxia. The Journal of Experimental Biology. 2001;204:3225-3234

[53] Looft JM, Herkert N, Frey-Law L. Modification of a three-compartment muscle fatigue model to predict peak torque decline during intermittent tasks. Journal of Biomechanics. 2018;77:16-25

[54] Clemente FM, Pracą GM, Bredt SDGT, Der Linden CMIV, Serra-Olivares J. External load variations between medium- and large-sided soccer games: Ball possession games vs regular games with small goals. Journal of Human Kinetics. 2019;70:191-198. DOI: 10.2478/hukin-2019-0031

[55] Walton CC, Keegan RJ, Martin M, Hallock H. The potential role for cognitive training in sport: More research needed. Frontiers in Psychology. 2018;9:1121. DOI: 10.3389/fpsyg.2018.01121

[56] Wiersma R, Stoter IK, Visscher C, Hettinga FJ, Elferink-Gemser MT. Development of 1500-m pacing behavior in junior speed skaters: A longitudinal study. International Journal of Sports Physiology and Performance. 2017;12:1224-1231. DOI: 10.1123/ijspp.2016-0517

[57] Whitehead AE, Jones HS, Williams EL, Rowley C, Quayle L, Marchant D, et al. Investigating the relationship between cognitions, pacing strategies and performance in 16.1 km cycling time trials using a think aloud protocol. Psychology of Sport and Exercise. 2018;34:95-109. DOI: 10.1016/j.psychsport.2017.10.001

[58] McGibbon KE, Pyne DB, Shephard ME, Thompson KG. Pacing in swimming: A systematic review. Sports Medicine. 2018;48:1621-1633. DOI: 10.1007/s40279-018-0901-9

[59] de Koning JJ, Foster C, Bakkum A, Kloppenburg S, Thiel C, Joseph T, et al. Regulation of pacing strategy during athletic competition. PLoS One. 2011;6:e15863. DOI: 10.1371/journal.pone.0015863

[60] Abbiss CR, Laursen PB. Describing and understanding pacing strategies during athletic competition. Sports Medicine. 2008;38:239-252. DOI: 10.2165/00007256-200838030-00004

[61] Ferraz R, Gonçalves B, Coutinho D, Marinho DA, Sampaio J, Marques MC. Pacing behaviour of players in team sports: Influence of match status manipulation and task duration knowledge. PLoS One. 2018;13: e0192399. DOI: 10.1371/journal. pone.0192399

[62] Edwards AM, Polman RCJ. Pacing and awareness: Brain regulation of physical activity. Sports Medicine. 2013;43:1057-1064. DOI: 10.1007/s40279-013-0091-4

[63] Smits BLMM, Pepping GJ, Hettinga FJ. Pacing and decision making in sport and exercise: The roles of perception and action in the regulation of exercise intensity. Sports Medicine. 2014;44:763-775. DOI: 10.1007/s40279-014-0163-0

[64] Stoter IK, Macintosh BR, Fletcher JR, Pootz S, Zijdewind I, Hettinga FJ. Pacing strategy, muscle fatigue, and technique in 1500-m speed-skating and cycling time trials. International Journal of Sports Physiology and Performance. 2016;11:337-343. DOI: 10.1123/ijspp.2014-0603

[65] Esteve-Lanao J, Lucia A, de Koning JJ, Foster C. How do humans control physiological strain during strenuous endurance exercise? PLoS One. 2008;3:e2943. DOI: 10.1371/journal.pone.0002943

[66] Baron B, Moullan F, Deruelle F, Noakes TD. The role of emotions on pacing strategies and performance in middle and long duration sport events. British Journal of Sports Medicine. 2011;45:511-517. DOI: 10.1136/bjsm.2009.059964

[67] Gibson ASC, Lambert EV, Rauch LHG, Tucker R, Baden DA, Foster C, et al. The role of information processing between the brain and peripheral physiological systems in pacing and perception of effort. Sports Medicine. 2006;36:705-722. DOI: 10.2165/00007256-200636080-00006

[68] Micklewright D, Kegerreis S, Raglin J, Hettinga F. Will the conscious–subconscious pacing quagmire help elucidate the mechanisms of self-paced exercise? New opportunities in dual process theory and process tracing methods. Sports Medicine. 2017;47: 1231-1239. DOI: 10.1007/s40279-016-0642-6

[69] Corbett J. An analysis of the pacing strategies adopted by elite athletes during track cycling. International Journal of Sports Physiology and Performance. 2009;4:195-205. DOI: 10.1123/ijspp.4.2.195

[70] Song J-H. Abandoning and modifying one action plan for alternatives. Philosophical Transactions of the Royal Society B: Biological Sciences. 2017;372:20160195. DOI: 10.1098/rstb.2016.0195

[71] Noorbergen OS, Konings MJ, Micklewright D, Elferink-Gemser MT, Hettinga FJ. Pacing behavior and tactical positioning in 500-and 1000-m short-track speed skating. International Journal of Sports Physiology and Performance. 2016;11:742-748. DOI: 10.1123/ijspp.2015-0384

[72] Konings MJ, Noorbergen OS, Parry D, Hettinga FJ. Pacing behavior and tactical positioning in 1500-m short-track speed skating. International Journal of Sports Physiology and Performance. 2016;11:122-129. DOI: 10.1123/ijspp.2015-0137

[73] Karageorghis CI, Priest DL. Music in the exercise domain: A review and synthesis (Part I). International Review of Sport and Exercise Psychology. 2012;5:44-66. DOI: 10.1080/1750984X.2011.631026

[74] Smits BLM, Polman RCJ, Otten B, Pepping G-J, Hettinga FJ. Cycling in the

absence of task-related feedback: Effects on pacing and performance. Frontiers in Physiology. 2016;7:348. DOI: 10.3389/fphys.2016.00348

[75] Teunissen LPJ, De Haan A, De Koning JJ, Daanen HAM. Effects of wind application on thermal perception and self-paced performance. European Journal of Applied Physiology. 2013;**113**:1705-1717. DOI: 10.1007/s00421-013-2596-9

[76] Dugas JP, Oosthuizen U, Tucker R, Noakes TD. Rates of fluid ingestion alter pacing but not thermoregulatory responses during prolonged exercise in hot and humid conditions with appropriate convective cooling. European Journal of Applied Physiology. 2009;**105**:69-80. DOI: 10.1007/s00421-008-0876-6

[77] Marcora SM, Staiano W, Manning V. Mental fatigue impairs physical performance in humans. Journal of Applied Physiology. 2009;**106**:857-864. DOI: 10.1152/japplphysiol.91324.2008

[78] Paterson S, Marino FE. Effect of deception of distance on prolonged cycling performance. Perceptual and Motor Skills. 2004;**98**:1017-1026. DOI: 10.2466/pms.98.3.1017-1026

[79] Ferraz R, Gonçalves B, Coutinho D, Oliveira R, Travassos B, Sampaio J, et al. Effects of knowing the task's duration on soccer players' positioning and pacing behaviour during small-sided games. International Journal of Environmental Research and Public Health. 2020;**17**:1-12. DOI: 10.3390/ijerph17113843

[80] RDA A, Silva-Cavalcante MD, Lima-Silva AE, Bertuzzi R. Fatigue development and perceived response during self-paced endurance exercise: State-of-the-art review. European Journal of Applied Physiology. 2021;**121**:687-696. DOI: 10.1007/s00421-020-04549-5

[81] Sakalidis KE, Burns J, Van Biesen D, Dreegia W, Hettinga FJ. The impact of cognitive functions and intellectual impairment on pacing and performance in sports. Psychology of Sport and Exercise. 2021;**52**:101840. DOI: 10.1016/j.psychsport.2020.101840

[82] Tee JC, Coopoo Y, Lambert M. Pacing characteristics of whole and part-game players in professional rugby union. European Journal of Sport Science. 2020;**20**:722-733. DOI: 10.1080/17461391.2019.1660410

[83] Pusenjak N, Grad A, Tusak M, Leskovsek M, Schwarzlin R. Can biofeedback training of psychophysiological responses enhance athletes' sport performance? A practitioner's perspective. The Physician and Sportsmedicine. 2015;**43**:287-299. DOI: 10.1080/00913847.2015.1069169

[84] Mauger AR, Jones AM, Williams CA. Influence of feedback and prior experience on pacing during a 4-km cycle time trial. Medicine and Science in Sports and Exercise. 2009;**41**:451-458. DOI: 10.1249/MSS.0b013e3181854957

[85] Faulkner J, Arnold T, Eston R. Effect of accurate and inaccurate distance feedback on performance markers and pacing strategies during running. Scandinavian Journal of Medicine & Science in Sports. 2011;**21**:e176-e183. DOI: 10.1111/j.1600-0838.2010.01233.x

[86] Marcora S. Perception of effort during exercise is independent of afferent feedback from skeletal muscles, heart, and lungs. Journal of Applied Physiology. 2009;**106**:2060-2062. DOI: 10.1152/japplphysiol.90378.2008

[87] Micklewright D, Papadopoulou E, Swart J, Noakes T. Previous experience influences pacing during 20 km time trial cycling. British Journal of Sports Medicine. 2010;**44**:952-960. DOI: 10.1136/bjsm.2009.057315

[88] Júnior JFCR, Mckenna Z, Amorim FT, Sena AFDC, Mendes TT, Veneroso CE, et al. Thermoregulatory and metabolic responses to a half-marathon run in hot, humid conditions. Journal of Thermal Biology. 2020;**93**:102734

[89] Negra Y, Chaabene H, Hammami M, Khlifa R, Gabbet T, Hachana Y. Allometric scaling and age related differences in change of direction speed performances of young soccer players. Science & Sports. 2016;**31**:19-26. DOI: 10.1016/j.scispo.2015.10.003

[90] Baden DA, McLean TL, Tucker R, Noakes TD, St.Clair GA. Effect of anticipation during unknown or unexpected exercise duration on rating of perceived exertion, affect, and physiological function. British Journal of Sports Medicine. 2005;**39**:742-746. DOI: 10.1136/bjsm.2004.016980

[91] Eston R, Stansfield R, Westoby P, Parfitt G. Effect of deception and expected exercise duration on psychological and physiological variables during treadmill running and cycling. Psychophysiology. 2012;**49**:462-469. DOI: 10.1111/j.1469-8986.2011.01330.x

[92] Clemente FM, Rabbani A, Ferreira R, Araújo JP. Drops in physical performance during intermittent small-sided and conditioned games in professional soccer players. Human Movement. 2020;**21**:7-14

[93] Batista J, Goncalves B, Sampaio J, Castro J, Abade E, Travassos B. The influence of coaches' instruction on technical actions, tactical behaviour, and external workload in football small-sided games. Montenegrin Journal of Sports Science and Medicine. 2019;**8**:29-36. DOI: 10.26773/mjssm.190305

[94] Branquinho L, Ferraz R, Marques MC. 5-a-side game as a tool for the coach in soccer training. Strength & Conditioning Journal. 2021;**43**:96-108. DOI: 10.1519/ssc.0000000000000629

[95] Ferraz R, van den Tillaar R, Marques MC. The effect of fatigue on kicking velocity in soccer players. Journal of Human Kinetics. 2012;**35**:97-107. DOI: 10.2478/v10078-012-0083-8

[96] Billaut F, Bishop DJ, Schaerz S, Noakes TD. Influence of knowledge of sprint number on pacing during repeated-sprint exercise. Medicine and Science in Sports and Exercise. 2011;**43**:665-672. DOI: 10.1249/MSS.0b013e3181f6ee3b

[97] Morton RH. Deception by manipulating the clock calibration influences cycle ergometer endurance time in males. Journal of Science and Medicine in Sport. 2009;**12**:332-337. DOI: 10.1016/j.jsams.2007.11.006

[98] Meeusen R, Watson P, Hasegawa H, Roelands B, Piacentini MF. Central fatigue: The serotonin hypothesis and beyond. Sports Medicine (Auckland, NZ). 2006;**36**:881-909. DOI: 10.2165/00007256-200636100-00006

[99] Enoka RM, Duchateau J. Translating fatigue to human performance. Medicine and Science in Sports and Exercise. 2016;**48**:2228-2238. DOI: 10.1249/MSS.0000000000000929

[100] Hureau TJ, Romer LM, Amann M. The 'sensory tolerance limit': A hypothetical construct determining exercise performance? European Journal of Sport Science. 2018;**18**:13-24. DOI: 10.1080/17461391.2016.1252428

[101] Davis L, Appleby R, Davis P, Wetherell M, Gustafsson H. The role of coach-athlete relationship quality in team sport athletes' psychophysiological exhaustion: Implications for physical and cognitive performance. Journal of Sports Sciences. 2018;**36**:1985-1992. DOI: 10.1080/02640414.2018.1429176

Chapter 2

The Importance of Sleep in Athletes

Júlio Costa, Pedro Figueiredo, Fábio Y. Nakamura and João Brito

Abstract

Sleep is an essential component for athletes' recovery from fatigue, due especially to its physiological and psychological restorative effects. Moreover, sleep is extremely important for numerous biological functions, and sleep deprivation can have significant effects on athletic performance in short-, medium-, and long term. For example, and considering the physiology of sleep for athletes, some hormonal responses that take place in the lead up to and during sleep (e.g., growth hormone—important role in muscle growth and repair) may be affected following exercise (i.e., training and competition), especially when compared with non-athlete's populations. Thus, monitoring sleep is also crucial to understand responses to training and readiness, enabling appropriate planning. Importantly, sleep monitoring also intends to reduce the risk of injury, illness, and nonfunctional overreaching. Moreover, an "individual approach" in athletes monitoring could help in better prescribe training contents and more adequately manage fatigue, as well as recommend pertinent post-match recovery strategies, such as sleep hygiene interventions. Overall, for understanding the athlete's sleep patterns/responses and to optimize the recovery strategies, it is crucial for comprehensive monitoring of his/her health, performance, fitness, and fatigue status.

Keywords: athletes, sleep interventions, sleep technology, performance, health

1. Introduction

Sleep is fundamental for sports performance, as well as for emotional regulation and development of the physical and mental health of athletes. In fact, inadequate sleep (e.g., reduced sleep duration and quality) may lead to an increased risk of injury and illness in athletes.

In recent years, growing interest in understanding the sleep of athletes has seen an increase in published studies [1]. In fact, athletes and coaches have ranked sleep as the most important recovery strategy [2]. Interestingly, the fundamental difference between recovery interventions with established protocols (e.g., cold water immersion, compression garments, electrical stimulation) [3] and sleeping lies in the fact that sleep initiation does not depend entirely on the willingness of the athlete [4].

During sleep, anabolic metabolism is upregulated [5], procedural memories are consolidated [6], and immune responses are augmented [7]. However, sleep loss or deprivation can have significant effects on performance, motivation, perception of effort, and cognition as well as numerous other biological functions [8]. Furthermore, sleep is associated with many physiological processes that may facilitate recovery from, and adaptation to, athletic training and competition [9]. Studies have analyzed the importance of sleep to regulate key molecular mechanisms (i.e., transcriptional regulatory proteins [10–12]), demonstrating that sleep has an integral role in metabolic homeostasis [13]. The capacity of humans to cope with physiological and psychological stressors is fundamental to athletic performance outcomes [14] and may be influenced by numerous factors, such as experience, fitness, motivation, and the normal fluctuation of physiological and behavioral procedures across a 24-h period (i.e., sleep–wake cycle, body temperature, hormone regulation) [15].

Importantly, the circadian rhythms are mainly controlled by the suprachiasmatic nucleus within the hypothalamus [16]. However, the suprachiasmatic nucleus is unable to continuously sustain control over these patterns (i.e., between the suprachiasmatic nucleus within the hypothalamus), as humans are extremely sensitive to changes in their normal environment [16, 17], most notably through the light–dark cycle [18]. When athletes face disturbances to their environments (e.g., training and/or competing close to bedtime sleep and travel), endogenous circadian rhythms and normal sleep-wake cycles can become desynchronized [16, 19]. These perturbations in sleeping patterns can cause an increase in homeostatic pressure and affect emotional regulation, core temperature, and circulating levels of melatonin, causing a delay in sleep onset [20].

Additionally, there is potential for sleep loss and neurocognitive and physiological performance to be compromised [9, 21–23]. Emerging research suggests that there are differences in sleep duration and quality between athletes and healthy controls. In contrast to non-athletes, athletes are often exposed to conditions that can interfere with sleep duration and quality, such as jet lag, unfamiliar sleeping environments, evening training, and/or competition and underlying fatigue [24].

In this sense, sleep monitoring has become a common practice in sport, and, in athletes, it may be useful to identify those who may need an intervention in terms of sleep disorders. Consequently, it is necessary to identify atypical patterns in the sleep and wakefulness of athletes and provide adequate sleep hygiene strategies to avoid disturbances in sleep duration and quality. Efficient and noninvasive methods and equipment, such as actigraphy and other alternatives to polysomnography, can provide detailed information about sleep and wakefulness during the sporting season.

Although there is high availability of information regarding the duration and quality of sleep in different age groups in the general population, information available in the scientific literature about sleep in athletes is still scarce. However, sleep is currently recognized as one of the essential components in the recovery from fatigue and, consequently, in the performance of athletes. Thus, it is essential that athletes, coaches, and clinicians understand the factors that can affect sleep, as well as realizing the usefulness of methods and equipment for assessing the duration and quality of sleep, as this process can result in better health and performance for the athlete.

2. The importance of sleep

Sleep is an essential component for athletes' recovery from fatigue, due especially to its physiological and psychological restorative effects [25]. In fact, it seems important that athletes learn to manage their sleeping and waking times, given the influence on circadian rhythm, since alterations in the biological clock may affect not only the duration and quality of sleep, but, mainly, sports performance [17].

Athletes and coaches recognize the importance of sleep as one of the most important strategies for recovering from fatigue and improving an athlete's performance [2]. However, during the competitive period, it is common for athletes to follow strict training and competition schedules, which, associated with intense training loads and the physical and emotional demands of competitions, may interfere and reduce the duration and quality of their sleep [26] and, consequently, decrease the fatigue recovery process [27]. This potential imbalance can actually occur when training and competitions are held close to bedtime [28]. Furthermore, exercise, when performed close to bedtime, may alter circadian rhythms [29] and sleep patterns (e.g., reducing sleep duration) [28, 30]. In fact, it seems important that athletes learn to manage their sleeping and waking times, given the influence on circadian rhythm, since alterations in the biological clock may affect not only the duration and quality of sleep, but, mainly, sports performance [2].

In the general population, less than 8 h of sleep per night may be associated with alterations in cognitive performance, mood, and wakefulness, as well as with increases in daytime sleepiness episodes [31]. This theme extends to younger athletes, who are expected to have a greater physiological need for sleep (8–10 h per night) compared with adults (7–9 h per night) and who often experience delays in sleep onset and awakening [32, 33]. Similarly, compared with adult athletes, young athletes have different daily commitments, such as school and social activities (including time spent online during the night), which can further alter sleep habits and/or wakefulness [34]. As an example, in an epidemiological study [35], significant reductions in neurocognitive performance (assessed through visual tests of memory and speed of response to a given visual stimulus) were observed in 7150 young athletes from different sports, who had a sleep duration of less than 5 h per night.

However, despite the high availability of information regarding the duration and quality of sleep in different age groups in the general population, in the scientific literature, the information available regarding the duration and quality of sleep in athletes is still scarce [36]. In fact, this seems contradictory given that sleep is currently recognized as one of the essential components in athletes' recovery [25]. Thus, there is a need to investigate, through sensitive and noninvasive methods, the monitoring of sleep patterns and wakefulness in athletes, in order to promote better sleep hygiene and, consequently, better recovery and performance.

3. Sleep, injuries, and performance

The current training and competition demands are topics with the greatest interest and discussion in the fields of sports science and sports medicine. This theme is commonly associated with the problem of sports injuries that affect athletes. In this sense, it is essential that clubs create ideal conditions for the training and development of athletes, integrating strategies and best practices for the prevention, treatment, and rehabilitation of injuries in an integrated perspective for athletes' health and performance.

Sleep can influence the risk of injury and illness. In a study of 122 athletes, it was observed that the risk of injury increased by 65% when athletes slept less than 8 h per night [37]. In another more recent study, it was possible to observe that 23 athletes with reduced sleep durations (<8 h) demonstrated a high association with the increase in musculoskeletal injuries. However, evidence in the literature is still very limited about this association. It is also important to note that sports injury is an emergent complex phenomenon, and the risk factors of injury comprise nonlinear associations between various factors such as the biomechanics, training and competitions workloads, as well as psychological and physiological characteristics. For example, according to Laux et al. [38] results, the highest risk for injury appears to occur from a synchronized growth in training and competitions workloads and loss in total sleep time; nonetheless, prospective randomized trials determining that decreased sleep quality leads an injury could require a more decisive response. Research on this topic may provide important information for coaches and practitioners in identifying potential strategies to maintain and improve athlete well-being.

Effects of inadequate sleep duration and quality on performance are likely to be seen specifically in competitive athletes, because of their high-performance demands being more likely to show the harmful effects of suboptimal sleep. Research studies have found negative results of sleep deficiency on athletic performance and well-being, specifically relative to time to exhaustion, muscle strength, and mood state [39, 40]. In a study of a sleep banking (i.e., sleep extension) for college basketball players ($n = 11$, 18–22 age), sleep duration was augmented by 110.9 ± 79.7 min ($p < 0.001$), together with significant increases in daytime sleepiness, reaction time, sprinting time, accuracy, fatigue, tension, depression, irritation, confusion, and mood disturbance [41]. In other study of cyclist's athletes and triathletes [42], an improved endurance performance was shown after three nights of sleep banking (~8.4 h sleep each night) compared with usual sleep (~6.8 h sleep each night), suggesting that endurance athletes' sleep must be >8 h each night to improve performance.

Considering the importance of examining sleep habits and wakefulness in athletes, the impact of training and competition schedules and loads on sleep indices has recently been explored [43–45]. In these studies, it was observed that sleep habits (i.e., the duration and quality of sleep) can be affected by schedule variations and by training and competition loads, especially when sessions are held at night, close to bedtime.

It should also be noted that the sleep habits and wakefulness of athletes may depend on the type of sport practiced [26]. For instance, Lastella et al. [26] investigated sleep/wake behavior of elite athletes, including young female and male athletes, and compared differences between athletes from individual (cycling, mountain bike, racewalking, swimming, and triathlon) and team sports (Australian football, basketball, soccer, and rugby union). Sleep/wake behaviors of elite athletes ($n = 124$) were well below the recommended 8 h of sleep per night, with shorter sleep duration existing in individual sports. These outcomes suggest that the amount of sleep the athletes obtain depends also on their sport.

That said, and although the duration and quality of an athlete's sleep may be associated with the schedules and loads of training and competition, it is also important to consider other factors that can influence sleep indices and wakefulness, namely age, sex, and chronotype [46]. For example, sex was identified as a risk factor for lifetime sleep problems in elite French athletes, with a greater incidence of sleep problems in female athletes [47]. Age has been shown to relate to the prevalence of poor sleep quality, with athletes >25 years of age reporting greater Pittsburgh Sleep Quality Index (PSQI) scores compared with ages <20 [48]; early fatherhood and/or motherhood could be a causal factor [49]. The age of the athletes was also classified as a risk factor for sleep disturbance previous to a competition; however, habitual sleep quality was not [50]. These findings may indicate that athletes who normally report good sleep quality are not necessarily resilient against sleep disturbance during, for instance, a major competition.

4. Measuring sleep

To detect and control sleep disorders, it is important to monitor sleep habits and perceptions of sleep through subjective and objective measures [51].

In general, the main recommendations on sleep monitoring point to polysomnography, which uses surface electrodes to monitor physiological parameters such as brain, muscle, cardiac, and respiratory activity [52]. Polysomnography is particularly useful for investigating sleep pathologies, including sleep-disordered breathing [53] and sleep disorders caused by concussion [54]. However, polysomnography is an expensive technique and requires specialized laboratory equipment, so its use in athletes in the real context is impractical [55].

On the other hand, actigraphy uses accelerometers placed in portable devices to record movements that, analyzed using algorithms, estimate the quality and duration of sleep [56]. Actigraphy is less expensive, noninvasive, and can be used in training and competition routines, ideally requiring two consecutive weeks of monitoring [57]. Thus, actigraphy emerges as the most accessible method to objectively monitor the sleep of athletes during the night [55]. Overall, wrist-worn accelerometers allow estimation of total sleep time (the total amount of sleep obtained during a sleep period), time in bed (the amount of time spent in bed attempting to sleep between bedtime and get-up time), wake up time (time at which a athlete got out of bed and stopped attempting to sleep), sleep onset time (transition from wakefulness into sleep), wake after sleep onset (number of min awake after sleep onset), latency (the period of time between bedtime and sleep onset time), and sleep efficiency (percentage of time in bed that was spent asleep) [55]. However, it is imperative to highlight that activity monitors tend to underestimate sleep in people who exhibit high levels of movement during light sleep [58]. In fact, some works showed that (elite) athletes obtain less sleep than the general population [59, 60] and present larger movement and fragmentation during sleep [61, 62]. Thus, and given the sleep characteristics of (elite) athletes, it is important to determine how well activity monitors are sensitive to recognize moments of sleep and vigilance in this type of population. This raises a potential issue with the use of activity monitors for measuring sleep in (elite) athletes.

Questionnaires and in particular "sleep diaries" are also used to record the start and end times for all sleep periods (i.e., night sleep and daily naps) [57]. Nevertheless, subjective reports (e.g., PSQI) might deviate from objective measures [63], especially with regard to mood and memory biases, while personality characteristics may also affect self-reported sleep ratings [64]. Indeed, some discrepancies have been detected when comparing subjective parameters with objective measures [65].

Additionally, and considering the ability of monitoring (objectively or subjectively) sleep duration and quality obtained by an (elite) athlete as a useful tool for evaluating recovery from training and competition [55], it is crucial to highlight the importance of individualized monitoring.

Although it is conventional to focus monitoring on group mean responses following a particular training intervention or competition, sport settings frequently produce diverse results with high and low responders being often lost in the averaged data reports [66, 67]. As a consequence, an increased attention for individualization of monitoring in sport settings has growth to a variety of athlete-monitoring approaches, allowing coaches to better manage fatigue and planning training prescription on an individual basis [68].

Nevertheless, research examining the sleep of athletes has typically averaged data across several nights, providing a mean estimate of usual sleep [26, 48, 61]. While such approaches are useful to allow basic insight into sleep (to better understand fatigue and recovery in athletes), they lack the sophistication to provide understanding of how sleep may vary across multiple nights at the individual level [69–71]. Moreover, individual variability can reflect differences within individuals over time [72], with high intra-individual variability in the athletes' sleep indicating the need for individualized sleep education strategies and interventions to promote appropriate sleep [69].

Although identifying the optimal amount of sleep on an individual basis may be difficult [73], young and adult athletes who exhibit average sleep of less than 8 or 7 h, respectively, likely warrant additional assessment to classify their sleep difficulties. Hence, those athletes that reveal deleterious effects of inadequate total sleep time should be stimulated to use sleep hygiene strategies to increase sleep during night and vigilance during the day [74]. Longitudinal monitoring of training and match load, sleep, fatigue (e.g., through heart rate variability), stress, and mood may not only help identify individuals at risk, but also monitor improvements in sleep, well-being, and performance after interventions [75].

Overall, it might be important to include sleep monitoring in (elite) athletes encompassing individual responses, in addition to group means [69]. Also, special attention should be given to the sleep behavior of (elite) athletes (e.g., total sleep time) during periods of congested fixtures, such as international competitions, since sleep deficits can impair performance [17], as already mention above (point 3).

5. Sleep hygiene

The implementation of strategies that promote sleep quality should be a priority for athletes. In fact, during sleep, fundamental physiological and psychological processes take place for the recovery from fatigue, so the optimization of sleep hygiene strategies increasingly assumes an important role in the routines and planning of those dedicated to improving sport performance.

A recent study [76] evaluated the effect of education on sleep hygiene in athletes. It was found that sleep hygiene education had a considerable positive impact on sleep indices. Educational programs on sleep hygiene in athletes provided a significant improvement in sleep duration and quality and reduced daytime sleepiness. Furthermore, research into the effects of sleep hygiene education on athletes, especially young people, is quite limited [31].

As mentioned before, there are several factors that can influence the duration and quality of sleep in athletes. Calendars congested with competitions and regular trips, competitions of great physical and emotional demand that take place at night, or constant changes in the morning time to wake up because of training and travel are examples of common factors that can negatively influence the duration and quality of sleep in athletes.

In this context, the management of light exposure emerges as fundamental, as this factor has a significant impact on sleep. Exposure to light influences the production of melatonin, so managing the times of exposure to artificial light throughout the day can be used as a sleep management and hygiene strategy. Additionally, in competitions that take place at night, athletes are exposed to immense artificial light: lighting in sports facilities, the projectors used by the media in interviews at the end of competitions, light from busses, airports, and planes.

On the other hand, social contexts may also be decisive. In recent studies carried out with female soccer players in Portugal, who usually start training very late, close to bedtime, due to their daily commitments (e.g., work, studies) that have to be reconciled with the training and match schedules, it was found that the athletes showed a reduction in total sleep time and length of time to fall asleep on training days performed at night, compared with training days performed during the day or on rest days (i.e., days without exercise) [28, 44]. It was pointed out that one of the additional explanations for the observed results could have been in the athletes' exposure to the light emitted in the stadium. In fact, these data are little studied in sport, but during the training days, the athletes were exposed to >1200 lux and 5600 K, with the bright polychromatic light ≥1000 lux, which could be enough to stimulate wakefulness effects during sleep [77]. However, it should be borne in mind that, currently, one of the main sources of exposure to light results from the use of electronic devices (especially smartphones and tablets) and that their use around bedtime is possibly the factor that most influences the sleep latency of athletes.

Thus, the term sleep hygiene, which refers to the recommendations, strategies, behaviors, and conditions developed to promote quality and duration of sleep, has been appearing more and more often in the list of sports planning tasks for athletes [25]. It is important to be aware that, unlike other possible recovery strategies used in sport (e.g., cryotherapy, massage, nutrition, nutritional supplementation), sleep has particularities that are not always controlled by the athlete themselves. Thus, bearing in mind the importance that sleep can have on sports performance, this is a subject that deserves the greatest attention of all those dedicated to promoting health and performance in athletes.

6. Conclusions

Athletes, coaches, and supporting staff should adopt a scientific approach to both designing and monitoring training programs. Appropriate health and load monitoring is crucial for determining whether a player is adapting to a training program and minimizing the risk of developing nonfunctional overreaching, illness, or injury. To gain understanding of the training and match demands and their effects on the player, several potential markers are available. However, very few of them have strong scientific evidence supporting their use. Moreover, it is important to note that athletes, from different types of sports, normally obtain inadequate sleep duration and quality. From an athletic point of view, reductions in performance, decision-making ability, learning, and cognition can occur alongside reductions in immune function and an increased susceptibility to injury gain.

In this respect, monitoring sleep in athletes can be useful for early detection and intervention before significant performance and health decrements are observed. Noninvasive and time-efficient methods/equipment such as wearable actigraphy monitors can provide detailed information about positive and negative adaptions over short and long periods throughout the competitive season. In addition, each athlete can perform the recordings at home and/or training facilities, adopting a "real world scenario" to grant high ecological validity to the research and/or practical interventions. The accumulated knowledge regarding the importance of sleep has sleep monitoring to become a popular strategy among (elite) athletes, coaches, and supporting staff. However, given the complexity of analyzing sleep patterns and the limited availability of athletes to participate in sleep studies, those indicators are yet poorly documented.

Overall, factors related to training and competition can alter sleep patterns in athletes. Therefore, topics such as: (1) sleep patterns and disorders among athletes; (2) sleep and optimal functioning among athletes; (3) screening, tracking, and assessment of athletes' sleep; and (4) interventions (i.e., sleep hygiene) to improve sleep must be further investigated.

Conflict of interest

The authors declare no conflict of interest.

Author details

Júlio Costa[1*], Pedro Figueiredo[1,2], Fábio Y. Nakamura[3] and João Brito[1]

1 Portugal Football School, Portuguese Football Federation (FPF), Oeiras, Portugal

2 CIDEFES, Universidade Lusófona, Lisboa, Portugal

3 Research Center in Sports Sciences, Health Sciences and Human Development (CIDESD), University of Maia (ISMAI), Maia, Portugal

*Address all correspondence to: jahdc@hotmail.com

IntechOpen

References

[1] Roberts SSH, Teo WP, Warmington SA. Effects of training and competition on the sleep of elite athletes: A systematic review and meta-analysis. British Journal of Sports Medicine. 2019;**53**(8):513-522. DOI: 10.1136/bjsports-2018-099322

[2] Fallon KE. Blood tests in tired elite athletes: Expectations of athletes, coaches and sport science/sports medicine staff. British Journal of Sports Medicine. 2007;**41**(1):41-44. DOI: 10.1136/bjsm.2006.030999

[3] Nedelec M, McCall A, Carling C, Legall F, Berthoin S, Dupont G. Recovery in soccer: Part ii: Recovery strategies. Sports Medicine. 2013;**43**(1):9-22. DOI: 10.1007/s40279-012-0002-0

[4] Nedelec M, Halson S, Abaidia AE, Ahmaidi S, Dupont G. Stress, sleep and recovery in elite soccer: A critical review of the literature. Sports Medicine. 2015;**45**(10):1387-1400. DOI: 10.1007/s40279-015-0358-z

[5] Chennaoui M, Arnal PJ, Drogou C, Sauvet F, Gomez-Merino D. Sleep extension increases IGF-I concentrations before and during sleep deprivation in healthy young men. Applied Physiology, Nutrition, and Metabolism. 2016;**41**(9):963-970. DOI: 10.1139/apnm-2016-0110

[6] Frank MG, Benington JH. The role of sleep in memory consolidation and brain plasticity: Dream or reality? The Neuroscientist. 2006;**12**(6):477-488. DOI: 10.1177/1073858406293552

[7] Besedovsky L, Lange T, Born J. Sleep and immune function. Pflügers Archiv: European Journal of Physiology. 2012;**463**(1):121-137. DOI: 10.1007/s00424-011-1044-0

[8] Halson SL. Sleep in elite athletes and nutritional interventions to enhance sleep. Sports Medicine. 2014;**44** (Suppl. 1):S13-S23

[9] Samuels C. Sleep, recovery, and performance: The new frontier in high-performance athletics. Neurologic Clinics. 2008;**26**(1):169-180. DOI: 10.1016/j.ncl.2007.11.012

[10] Allada R, Siegel JM. Unearthing the phylogenetic roots of sleep. Current Biology. 2008;**18**(15):R670-R6R9

[11] Crocker A, Sehgal A. Genetic analysis of sleep. Genes & Development. 2010;**24**(12):1220-1235

[12] Abel T, Havekes R, Saletin JM, Walker MP. Sleep, plasticity and memory from molecules to whole-brain networks. Current Biology. 2013;**23**(17):R774-R788

[13] Xie L, Kang H, Xu Q, Chen MJ, Liao Y, Thiyagarajan M, et al. Sleep drives metabolite clearance from the adult brain. Science. 2013;**342**(6156):373-377

[14] Bishop D. An applied research model for the sport sciences. Sports Medicine. 2008;**38**(3):253-263

[15] Drust B, Waterhouse J, Atkinson G, Edwards B, Reilly T. Circadian rhythms in sports performance: An update. Chronobiology International. 2005;**22**(1):21-44

[16] Beersma DG, Gordijn MC. Circadian control of the sleep-wake cycle. Physiology & Behavior. 2007;**90**(2-3):190-195

[17] Fullagar H, Skorski S, Duffield R, Hammes D, Coutts AJ, Meyer T. Sleep and athletic performance: The effects of sleep loss on exercise performance, and physiological and cognitive responses to exercise. Sports Medicine. 2015;**45**(2):161-186. DOI: 10.1007/s40279-014-0260-0

[18] Czeisler CA, Allan JS, Strogatz SH, Ronda JM, Sanchez R, Rios CD, et al. Bright light resets the human circadian pacemaker independent of the timing of the sleep-wake cycle. Science. 1986;**233**(4764):667-671

[19] Reilly T, Edwards B. Altered sleep-wake cycles and physical performance in athletes. Physiological and Behavior. 2007;**90**(2-3):274-284. DOI: 10.1016/j. physbeh.2006.09.017

[20] Lack LC, Wright HR. Chronobiology of sleep in humans. Cellular and Molecular Life Sciences. 2007;**64**(10):1205-1215

[21] Halson SL. Nutrition, sleep and recovery. European Journal of Sport Science. 2008;**8**(2):119-126

[22] Goel N, Rao H, Durmer JS, Dinges DF. Neurocognitive consequences of sleep deprivation. Seminars in Neurology. 2009;**29**(4):320-339

[23] Banks S, Dinges DF. Behavioral and physiological consequences of sleep restriction. Journal of Clinical of Sleep Medicine. 2007;**3**(5):519-528

[24] Robey E, Dawson B, Halson S, Gregson W, Goodman C, Eastwood P. Sleep quantity and quality in elite youth soccer players: A pilot study. European Journal of Sport Science. 2014;**14**(5): 410-417. DOI: 10.1080/17461391.2013. 843024

[25] Walsh NP, Halson SL, Sargent C, Roach GD, Nedelec M, Gupta L, et al. Sleep and the athlete: Narrative review and 2021 expert consensus recommendations. British Journal of Sports Medicine. 2020;**55**(7):356-368. DOI: 10.1136/bjsports-2020-102025

[26] Lastella M, Roach GD, Halson SL, Sargent C. Sleep/wake behaviours of elite athletes from individual and team sports. European Journal of Sport Science. 2015;**15**(2):94-100

[27] Vitale JA, Banfi G, Galbiati A, Ferini-Strambi L, Torre A. Effect of night-game on actigraphy-based sleep quality and perceived recovery in top-level volleyball athletes. International Journal of Sports Physiology and Performance. 2018;**14**: 1-14. DOI: 10.1123/ijspp.2018-0194

[28] Costa JA, Brito J, Nakamura FY, Oliveira EM, Costa OP, Rebelo AN. Does night-training load affect sleep patterns and nocturnal cardiac autonomic activity in high-level female soccer players? International Journal of Sports Physiology and Performance. 2018;**14**(6):779-787. DOI: 10.1123/ ijspp.2018-0652

[29] Buman MP, Phillips BA, Youngstedt SD, Kline CE, Hirshkowitz M. Does nighttime exercise really disturb sleep? Results from the 2013 National Sleep Foundation Sleep in America Poll. Sleep Medicine. 2014;**15**(7):755-761. DOI: 10.1016/j.sleep.2014.01.008

[30] Fowler P, Duffield R, Vaile J. Effects of simulated domestic and international air travel on sleep, performance, and recovery for team sports. Scandinavian Journal of Medicine & Science in Sports. 2015;**25**(3):441-451. DOI: 10.1111/ sms.12227

[31] Fox JL, Scanlan AT, Stanton R, Sargent C. Insufficient sleep in young athletes? Causes, consequences, and potential treatments. Sports Medicine. 2020;**50**(3):461-470

[32] Crowley S, Carskadon M. Modifications to weekend recovery sleep delay circadian phase in older adolescents. Chronobiology International. 2010;**27**(7):1469-1492. DOI: 10.3109/07420528.2010.503293

[33] Hirshkowitz M, Whiton K, Albert S, Alessi C, Bruni O, DonCarlos L, et al. National Sleep Foundation's sleep time duration recommendations: Methodology and

results summary. Sleep Health. 2015;**1**(1):40-43. DOI: 10.1016/j. sleh.2014.12.010

[34] Carskadon M. Sleep in adolescents: The perfect storm. Pediatric Clinics of North America. 2011;**58**(3):637-647. DOI: 10.1016/j.pcl.2011.03.003

[35] Sufrinko A, Johnson EW, Henry LC. The influence of sleep duration and sleep-related symptoms on baseline neurocognitive performance among male and female high school athletes. Neuropsychology. 2016;**30**(4):484-491

[36] Sargent C, Lastella M, Halson SL, Roach GD. How much sleep does an elite athlete need? International Journal of Sports Physiology and Performance. 2021;**16**:1-12

[37] Milewski MD, Skaggs DL, Bishop GA, Pace JL, Ibrahim DA, Wren TA, et al. Chronic lack of sleep is associated with increased sports injuries in adolescent athletes. Journal of Pediatric Orthopedics. 2014;**34**(2):129-133. DOI: 10.1097/BPO.00000000000 00151

[38] Laux P, Krumm B, Diers M, Flor H. Recovery-stress balance and injury risk in professional football players: A prospective study. Journal of Sports Sciences. 2015;**33**(20):2140-2148. DOI: 10.1080/02640414.2015.1064538

[39] VanHelder T, Radomski MW. Sleep deprivation and the effect on exercise performance. Sports Medicine. 1989;**7**(4):235-247.DOI:10.2165/00007256-198907040-00002

[40] Van Ryswyk E, Weeks R, Bandick L, O'Keefe M, Vakulin A, Catcheside P, et al. A novel sleep optimisation programme to improve athletes' well-being and performance. European Journal of Sport Science. 2017;**17**(2):144-151. DOI: 10.1080/ 17461391.2016.1221470

[41] Mah CD, Mah KE, Kezirian EJ, Dement WC. The effects of sleep extension on the athletic performance of collegiate basketball players. Sleep. 2011;**34**(7):943-950. DOI: 10.5665/ SLEEP.1132

[42] Roberts SSH, Teo WP, Aisbett B, Warmington SA. Extended sleep maintains endurance performance better than normal or restricted sleep. Medicine and Science in Sports and Exercise. 2019;**51**(12):2516-2523. DOI: 10.1249/MSS.000000000000 2071

[43] Lastella M, Roach GD, Vincent GE, Scanlan AT, Halson SL, Sargent C. The impact of training load on sleep during a 14-day training camp in elite, adolescent, female basketball players. International Journal of Sports Physiology and Performance. 2020;**15**:1-7. DOI: 10.1123/ijspp.2019-0157

[44] Costa JA, Brito J, Nakamura FY, Figueiredo P, Oliveira E, Rebelo A. Sleep patterns and nocturnal cardiac autonomic activity in female athletes are affected by the timing of exercise and match location. Chronobiology International. 2018;**36**(3):360-373. DOI: 10.1080/07420528.2018.1545782

[45] Figueiredo P, Costa J, Lastella M, Morais J, Brito J. Sleep indices and cardiac autonomic activity responses during an international tournament in a youth national soccer team. International Journal of Environmental Research and Public Health. 2021;**18**(4): 2076. DOI: 10.3390/ijerph18042076

[46] Lastella M, Roach GD, Halson SL, Sargent C. The chronotype of elite athletes. The Journal of Human Kinetics. 2016;**54**:219-225. Epub 2016/12/30. DOI: 10.1515/hukin-2016-0049

[47] Schaal K, Tafflet M, Nassif H, Thibault V, Pichard C, Alcotte M, et al. Psychological balance in high level athletes: Gender-based differences and sport-specific patterns. PLoS One.

2011;**6**(5):e19007. Epub 2011/05/17. DOI: 10.1371/journal.pone.0019007

[48] Swinbourne R, Gill N, Vaile J, Smart D. Prevalence of poor sleep quality, sleepiness and obstructive sleep apnoea risk factors in athletes. European Journal of Sport Science. 2016;**16**(7): 850-858. Epub 2015/12/25. DOI: 10.1080/17461391.2015.1120781

[49] Finan PH, Quartana PJ, Smith MT. The effects of sleep continuity disruption on positive mood and sleep architecture in healthy adults. Sleep. 2015;**38**(11):1735-1742. DOI: 10.5665/sleep.5154

[50] Juliff LE, Halson SL, Peiffer JJ. Understanding sleep disturbance in athletes prior to important competitions. Journal of Science and Medicine in Sport. 2015;**18**(1):13-18. DOI: 10.1016/j.jsams.2014.02.007

[51] Myllymaki T, Kyrolainen H, Savolainen K, Hokka L, Jakonen R, Juuti T, et al. Effects of vigorous late-night exercise on sleep quality and cardiac autonomic activity. Journal of Sleep Research. 2011;**20**(1 Pt 2):146-153

[52] Kushida CA, Littner MR, Morgenthaler T, Alessi CA, Bailey D, Coleman J Jr, et al. Practice parameters for the indications for polysomnography and related procedures: An update for 2005. Sleep. 2005;**28**(4):499-521

[53] George CF, Kab V, Levy AM. Increased prevalence of sleep-disordered breathing among professional football players. The New England Journal of Medicine. 2003; **348**(4):367-368. DOI: 10.1056/NEJM200301233480422

[54] Gosselin N, Lassonde M, Petit D, Leclerc S, Mongrain V, Collie A, et al. Sleep following sport-related concussions. Sleep Medicine. 2009;**10**(1):35-46. DOI: 10.1016/j.sleep.2007.11.023

[55] Sargent C, Lastella M, Halson SL, Roach GD. The validity of activity monitors for measuring sleep in elite athletes. Journal of Science and Medicine in Sport. 2016;**19**(10):848-853. DOI: 10.1016/j.jsams.2015.12.007

[56] Ancoli-Israel S, Cole R, Alessi C, Chambers M, Moorcroft W, Pollak CP. The role of actigraphy in the study of sleep and circadian rhythms. Sleep. 2003;**26**(3):342-392

[57] Halson SL. Sleep monitoring in athletes: Motivation, methods, miscalculations and why it matters. Sports Medicine. 2019;**49**:1487-1497. DOI: 10.1007/s40279-019-01119-4

[58] Tryon WW. Issues of validity in actigraphic sleep assessment. Sleep. 2004;**27**(1):158-165

[59] Roach GD, Schmidt WF, Aughey RJ, Bourdon PC, Soria R, Claros JC, et al. The sleep of elite athletes at sea level and high altitude: A comparison of sea-level natives and high-altitude natives (ISA3600). British Journal of Sports Medicine. 2013;**47**(Suppl. 1):i114-i120. DOI: 10.1136/bjsports-2013-092843

[60] Sargent C, Halson S, Roach GD. Sleep or swim? Early-morning training severely restricts the amount of sleep obtained by elite swimmers. European Journal of Sport Science. 2014;**14** (Suppl 1):S310-S315. DOI: 10.1080/17461391.2012.696711

[61] Leeder J, Glaister M, Pizzoferro K, Dawson J, Pedlar C. Sleep duration and quality in elite athletes measured using wristwatch actigraphy. Journal of Sports Science. 2012;**30**(6):541-545

[62] Taylor SR, Rogers GG, Driver HS. Effects of training volume on sleep, psychological, and selected physiological profiles of elite female swimmers. Medicine & Science in Sports & Exercise. 1997;**29**(5):688-693

[63] Kawada T. Agreement rates for sleep/wake judgments obtained via accelerometer and sleep diary: A comparison. Behavior Research Methods. 2008;**40**(4):1026-1029

[64] Jackowska M, Dockray S, Hendrickx H, Steptoe A. Psychosocial factors and sleep efficiency: Discrepancies between subjective and objective evaluations of sleep. Psychosomatic Medicine. 2011; **73**(9):810-816

[65] Kolling S, Endler S, Ferrauti A, Meyer T, Kellmann M. Comparing subjective with objective sleep parameters via multisensory actigraphy in german physical education students. Behavioral Sleep Medicine. 2016;**14**(4): 389-405

[66] Mann TN, Lamberts RP, Lambert MI. High responders and low responders: Factors associated with individual variation in response to standardized training. Sports Medicine. 2014;**44**(8):1113-1124. DOI: 10.1007/ s40279-014-0197-3

[67] Chrzanowski-Smith OJ, Piatrikova E, Betts JA, Williams S, Gonzalez JT. Variability in exercise physiology: Can capturing intra-individual variation help better understand true inter-individual responses? European Journal of Applied Physiology. 2019;**20**:452-460. DOI: 10.1080/17461391.2019.1655100

[68] Halson SL. Monitoring training load to understand fatigue in athletes. Sports Medicine. 2014;**44**(Suppl 2):S139-S147. DOI: 10.1007/s40279-014-0253-z

[69] Nedelec M, Aloulou A, Duforez F, Meyer T, Dupont G. The variability of sleep among elite athletes. Sports Medicine—Open. 2018;**4**(1):34. DOI: 10.1186/s40798-018-0151-2

[70] Buchheit M. Monitoring training status with HR measures: Do all roads lead to Rome? Frontiers in Physiology. 2014;**5**:73. DOI: 10.3389/fphys.2014. 00073

[71] Bei B, Wiley JF, Trinder J, Manber R. Beyond the mean: A systematic review on the correlates of daily intraindividual variability of sleep/wake patterns. Sleep Medicine Reviews. 2016;**28**:108-124. DOI: 10.1016/j.smrv.2015.06.003

[72] Van Dongen HP. Analysis of inter- and intra-individual variability. Journal of Sleep Research. 2005;**14**(2):201-203. DOI: 10.1111/j.1365-2869.2005.00453.x

[73] Watson AM. Sleep and athletic performance. Current Sports Medicine Reports. 2017;**16**(6):413-418. DOI: 10.1249/JSR.0000000000000418

[74] Gupta L, Morgan K, Gilchrist S. Does elite sport degrade sleep quality? A systematic review. Sports Medicine. 2017;**47**(7):1317-1333. DOI: 10.1007/ s40279-016-0650-6

[75] Simpson NS, Gibbs EL, Matheson GO. Optimizing sleep to maximize performance: Implications and recommendations for elite athletes. Scandinavian Journal of Medicine & Science in Sports. 2017;**27**(3):266-274. DOI: 10.1111/sms.12703

[76] Driller MW, Lastella M, Sharp AP. Individualized sleep education improves subjective and objective sleep indices in elite cricket athletes: A pilot study. Journal of Sports Sciences. 2019;**37**(17): 2121-2125. DOI: 10.1080/02640414.2019. 1616900

[77] Cajochen C. Alerting effects of light. Sleep Medicine Reviews. 2007;**11**(6):453-464. DOI: 10.1016/j. smrv.2007.07.009

Chapter 3

Physiological Adaptions to Acute Hypoxia

Erich Hohenauer

Abstract

When tissues are insufficiently supplied with oxygen, the environment is said to be hypoxic. Acute (exposures to) hypoxia can occur occupationally, within the scope of training and competitions or under pathological conditions. The increasing interest in acute exposure to altitude for training and research purposes makes it more important than ever to understand the physiological processes that occur under hypoxic conditions. Therefore, the scope of this chapter is to describe the main types of hypoxia on the oxygen cascade, to summarize the physiological consequences of acute hypoxia on the three main areas and to highlight the clinical consequences of acute hypoxia exposures for healthcare practitioners.

Keywords: hypoxia, altitude, oxygen, physiology, cardiorespiratory

1. Introduction

The human body cells need the energy to maintain their functions. This energy is mainly provided by sugar, carbohydrates and fat. To utilize these nutritive substances and to produce energy in return, inspired oxygen (O_2) from the air is needed. In the mitochondrial electron transport chain, O_2 is the final electron acceptor to generate ATP within the eukaryotic cells [1]. Whilst O_2 is needed for most life on earth, most of the earth's atmosphere does not contain a lot of O_2. From the surface of the planet, up to the border of space, the atmosphere contains a constant fraction of around 21% O_2 (often expressed as the F_iO_2 of around 0.21), 78% of nitrogen, 0.9% argon and 0.1% of other gases like carbon dioxide, methane, water vapor, etc. At sea level, the partial pressure of the above-mentioned gases can be estimated to be 593 mmHg for nitrogen, 160 mmHg for oxygen and 7.6 mmHg for argon. Indeed, the weight of air is responsible for atmospheric pressure.

It's well known that increasing altitude leads to quasi-exponential reductions in barometric pressure (P_B). At the summit of Mt. Everest (8848 m), the P_B is about one-third of the sea-level values. The reduced atmospheric pressure has therefore a direct influence on the partial pressure of inspired oxygen, which can be seen in **Figure 1**.

The inspired partial pressure of oxygen (P_iO_2) is lower than atmospheric oxygen partial pressure because water vapor is in the airways. The pressure of water vapor (PH_2O), which is not dependent on atmospheric pressure but temperature, should be taken into account when P_iO_2 is calculated [2]. The inhaled air gases will get humified and warmed by the airways and as a result, the PH_2O will adjust the partial pressure of all inhaled gases, including O_2.

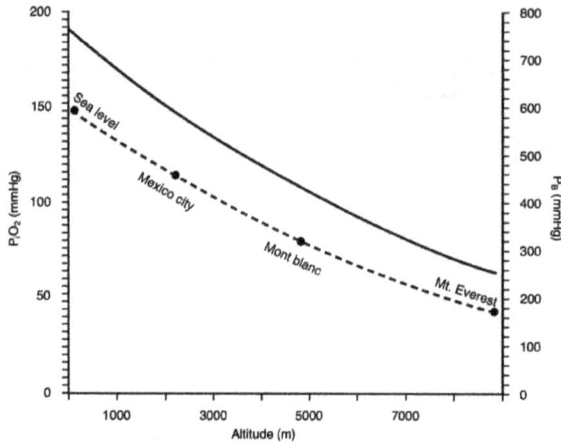

Figure 1.
Relationship between barometric pressure (P_B), partial pressure of inspired oxygen (P_iO_2) and altitude. P_B and P_iO_2 decrease exponentially with increasing altitude at a constant F_iO_2 of 21%. The solid line represents P_B and the broken line represents P_iO_2.

Accordingly, the product of P_iO_2 can be calculated using Eq. (1):

$$P_iO_2 = (P_B - PH_2O) \times F_iO_2. \tag{1}$$

Since P_B is known to be approximately 760 mmHg at sea level, PH_2O is normally about 47 mmHg and O_2 makes up to 20.93% (F_iO_2 of 0.2093), P_iO_2 is equal to 0.20932 multiplied by 713 mmHg.

Consequently, hypoxia is defined as a combination of P_B and the F_iO_2 that results in any P_iO_2 under a normoxic value of 150 mmHg [3]. However, the duration of hypoxic exposures as well as the magnitude of P_B reductions has a significant impact on the (patho-)physiological response. Examples of fast-changing normoxic to hypoxic environments are fast ascended on the mountain summits during mountaineering, military and rescue services and travels with fast transportation to altitude. Acute mountain sickness is well-known to occur due to extensive and fast decreases in P_b, normally beginning at an altitude of above 2500 m. The Lake Louis Consensus Group defined acute mountain sickness as the presence of headache in an unacclimatised person (recently arriving at an altitude above 2500 m), plus the presence of one or more of the following symptoms: gastrointestinal symptoms, fatigue and/or weakness, dizziness or a positive clinical functional score, resulting in a total score of ≥ 3 [4]. If not treated correctly, people with acute mountain sickness can develop high-altitude pulmonary oedema or high-altitude cerebral oedema [5]. However, if the human body is gradually exposed to hypoxic conditions, it can acclimatize and adapt.

The following chapters will focus on the main types of hypoxia, the physiological consequences of acute hypoxia and the clinical consequences of the current chapter.

2. Types of hypoxia

Insufficient O_2 supply to the human tissues can have various reasons and can lead to severely impaired body functions. There are four main types of hypoxia, which can be classified as hypoxaemic hypoxia, anemic hypoxia, stagnant hypoxia and histotoxic hypoxia.

2.1 Hypoxaemic type

One of the most common types of hypoxia is called generalized or hypoxic hypoxia, which is generated from the actual (natural/simulated) environment and inside the lungs. This type is caused by a reduction of the partial pressure of alveolar O_2 (P_AO_2) [6]. This value is well known and a great help to calculate the partial pressure of oxygen inside the alveoli (as it is not possible to collect gases directly from the alveoli), which can be used for potential cell diffusion [7]. The alveolar gas equation uses three variables to calculate the alveolar concentration of oxygen, which can be seen in Eq. (2):

$$P_AO_2 = F_iO_2 \times (P_B - PH_2O) - P_aCO_2 / RQ \tag{2}$$

where P_aCO_2 is the partial pressure of carbon dioxide which is under normal physiological conditions approximately 40 mmHg. RQ is the respiratory quotient which is, the ratio of the volume of produced CO_2 divided by the volume of consumed O_2 during the same time [8]. Dependent on metabolic activity and diet, RQ is considered to be around 0.825 [9], within a physiological range between 0.70 and 1.00. Consequently, P_AO_2 at sea level is: $0.2093 \times (760-47) - 40/0.825 = 100.7$ mmHg. P_AO_2 is the main driving factor for alveolar diffusion and thus O_2 supply on a cellular level.

Hypoxic hypoxia can be observed typically when F_iO_2 is low, during hypoventilation of the lungs or at the presence of pathological airway conditions. Low F_iO_2 levels can occur due to failure of gas delivery systems, inadequate supply from altitude simulating machines, or e.g., exorbitant inhalation of nitrous oxide during anesthesia [10]. Hypoventilation can occur due to insufficient respiratory rate, obstruction of airways, skeletal deformities, respiratory muscle paralysis, etc. Severe lung diseases (e.g., pulmonary fibrosis, pulmonary embolism) can also lead to alveolar-capillary diffusion blockade [11]. Hypoxic hypoxia affects the entire body. Typical symptoms are agitation and anxiety while low blood O_2 goes along with increased heart rate, dyspnea and bluish color of the skin.

2.2 Anemic type

Anemic hypoxia is caused by reduced oxygen transport capacity in the blood [12]. The red blood cells (erythrocytes) are responsible for the transport of O_2 through the body [13]. Around 90% of the erythrocyte is made up of haemoglobin, the iron-containing protein that binds O_2 on its heme. Although, the arterial oxygen tension is normal at this type, reduced erythrocytes/haemoglobin or functional insufficiency of haemoglobin leads to impaired oxygen delivery to the tissues [14].

A deficiency in the number of erythrocytes can result, for example, from excessive blood loss after trauma. Other forms of the reduced number of erythrocytes can be present in case of abnormal red blood cell breakdown (haemolytic anemia) [15]. Increased haemolysis can be observed during hereditary spherocytosis, sickle cell disease or autoimmune diseases (e.g., aplastic anemia) [16].

Deficiencies of different factors can also lead to severe anemia. Iron is the main component of haemoglobin, giving the blood the red color and is the prime carrier of oxygen. During the physiological haemolysis, iron will be bound to the glycoprotein transferrin for transportation to the bone marrow, where it will be reused for haemoglobin synthesis. This process helps to limit an extensive loss of iron from the body. However, iron deficiency is one of the main causes of anemia, called microcytic hypochromic anemia [6]. This type of anemia can be caused by any factor which reduces the body's iron storage, leading to small erythrocytes with reduced haemoglobin mass [17]. In contrast, deficiencies in vitamin B_{12} or folic acid can cause anemia due to abnormally enlarged erythrocytes and their immature precursors, called macrocytic hyperchromic anemia [18].

Functional insufficiency of haemoglobin is associated with reduced oxygen binding capacity. An example is an intoxication through excessive carbon monoxide inhalation. Compared to oxygen, carbon monoxide has a 200–300 times higher affinity to haemoglobin. After inhalation, carbon monoxide reaches the respiratory gas exchange zone and binds on haemoglobin [10]. This chemical binding process leads to the formation of carboxyhaemoglobin. Consequently, oxygen-carrying capacity is decreased which will lead to reduced oxygen transportation to the tissues and as a consequence tissue hypoxia [19]. Another possibility of functional insufficiency for the transportation of oxygen is methaemoglobinemia. Haemoglobin changes to methaemoglobin, when bivalent iron (Fe^{2+}) is oxidized to Fe^{3+}, which is worthless for oxygen transport [20]. Under normal circumstances, methaemoglobin reductase limits the build-up of methaemoglobin through the reduction of haemoglobin oxidation [21]. Patients with a deficiency of methaemoglobin reductase, strong oxidative stress (e.g., smoking) and medication can therefor experience very low concentrations of tissue oxygenation, demonstrating comparable symptoms as seen in hypoxic hypoxia. However, it must be mentioned, that the unfavorable conditions of low tissue O_2 can be compensated better during hypoxic hypoxia than during anemic hypoxia.

2.3 Stagnant type

Stagnant, also called ischemic or circulatory hypoxia takes place as a cause of insufficient blood supply to the tissues while the blood is normally oxygenated. Ischemic hypoxia can be observed on a central and local level [6].

Central circulatory hypoxia can often be observed in patients with cardiac manifestations. If the left ventricular output is for example decreased, blood flow to the organs is impaired [12]. This can also happen during shock or, at a local level after strong vasoconstriction (e.g., cold exposures) or venous stagnation of blood [22]. Oxygen can only be stored to the very limited amount within the human cells. Even myoglobin, binding O_2 on its heme protein, has a very limited oxygen storage capacity [23]. Consequently, myoglobin is more involved in transportation than the storage of oxygen. Oxygen saturated myoglobin enables facilitated intercellular O_2 transportation, because the oxygen-enriched myoglobin molecules can "move" within the cells (facilitated diffusion) which is extremely important at a low partial pressure of O_2 (PO_2) [24]. Although, the gas exchange rate on the alveolar level, the concentration of haemoglobin, oxygen content and tension are on a normal level, O_2 extraction at the level of the capillaries will be increased [6]. This process will directly elevate the arteriovenous difference of blood O_2 content leading to venous hypoxia. However, as the increased oxygen extraction is normally insufficient to supply the tissue with an adequate amount of O_2, this process will lead to impaired cellular oxygen coverage and impaired functioning.

2.4 Histotoxic type

Histotoxic hypoxia or dysoxia is a state, where cells are unable to utilize oxygen effectively [12]. This is the case, when the mitochondrial terminal oxidation is disturbed while there is sufficient oxygen available in the blood. Dysoxia will therefore lead to a pathological reduction in ATP production by the mitochondria and is not preceded by hypoxaemia [6].

An example of histotoxic hypoxia is the intoxication with cyanides, which can occur from fire sources. Intravenous and inhalation of cyanide produce a more rapid onset of hypoxia than the oral or transdermal route due to the fast diffusion into the bloodstream [25]. The main effect of cyanide intoxication is related to the inhibition of oxidative phosphorylation, where oxygen is utilized for ATP production. Cyanide can reversibly bind to the enzyme cytochrome C oxidase, blocking the mitochondrial transport chain. This will cause cellular hypoxia and, as mentioned above, pathological low levels of ATP, causing metabolic acidosis and impairment of vital functions [26, 27].

3. Physiological consequences of acute hypoxia

Rapid ascends from sea level to altitude and sudden exposure to a hypoxic environment will immediately lead to acute physiological responses to adapt to the acute hypoxaemic situation [28]. The degree of acute hypoxic stress about time can lead to symptoms ranging from dizziness, feeling of unreality and dim visions to rapid unconsciousness [29]. Sudden exposure to the summit of Mt. Everest will for example lead to unconsciousness within 2 min. However, when the same amount of hypoxaemia is experienced over several days to weeks, one could function relatively well under these conditions. This adjustment is called acclimatization which is a complex process over time and shows great variability within individuals [29]. In the following chapters, the acute response to sudden exposure to a hypoxic environment is discussed.

3.1 Respiratory system

The respiratory system will directly respond to the low oxygen availability in the air and is often seen as the primary defense against the hypoxic environment. Chemosensory systems will rapidly lead to increased pulmonary ventilation because of compromised O_2 availability [30]. These regulatory responses can be attributed due to specialized chemoreceptors such as the carotid bodies in the arterial circulation and neuroepithelial bodies in the respiratory tract as well as the direct response of vascular smooth muscles to hypoxia [31].

Whilst hypoxia acts as a vasodilator in the systemic circulation, it has been observed, that the vessels of the pulmonary vasculature constrict under hypoxia, leading to pulmonary hypertension [32, 33]. Hypoxic vasoconstriction is intrinsic to the pulmonary vasculature smooth muscle cells and is initiated by the inhibition of K^+ channels which set the membrane potential [34]. This process will lead to depolarization, activation of Ca^{2+} channels as a result of the electrical impulse and, as a consequence, an increase in cytosolic calcium levels and therefore constriction of the myocytes [31]. Pulmonary hypertension might help to match ventilation and perfusion within the lungs. However, pulmonary hypertension can also lead to severe pathological situations (e.g., altitude-related right heart failure).

Carotid bodies, sensitive to monitoring a drop in arterial O_2 levels, and neuroepithelial bodies, detecting changes in inspired O_2, respond immediately to decreased O_2 supply [35]. Both respond by activating efferent chemosensory fibers to produce cardiorespiratory adjustments during hypoxic exposures [36, 37]. When low arterial PO_2 is detected, the carotid body signals the central respiratory center to increase the (minute) ventilation. The increased ventilation of the respiratory tract can be primarily associated with an elevated tidal volume and an even greater elevation in respiratory rate [38]. This hypoxic ventilatory response counteracts the hypoxic environment by decreasing P_ACO_2, increasing P_AO_2 and therefore improving oxygen delivery. Genetical determinants, as well as various external factors (metabolic and respiratory stimulants), lead to wide inter-individual variety of ventilatory response intensity [39]. The increased ventilatory response demonstrates that adaptive processes are taking place and a "good" ventilatory response is known to enhance acclimatization and performance and that a very low response may contribute to the formation of illness [39, 40]. However, hyperventilation will subsequently lead to hypocapnia (increased pH) known as respiratory alkalosis by reducing the amount of carbon dioxide in the alveoli [41]. This condition will cause the oxygen dissociation curve to shift to the left and to further keep respiratory ventilation high. However, hypocapnia will also counteract the central respiratory center activation and thus limit further ventilatory increases [40, 42]. On the other hand, to reduce respiratory alkalosis, more bicarbonate will be produced from the kidneys to decrease the pH toward normal levels. This means that pulmonary ventilation is driven by low arterial PO_2 and limited due to hypocapnia-induced alkalosis at the same time. This becomes clear when looking into Eq. (3), defining the alveolar ventilation as follows:

$$V_A = 0.863 \times VCO_2 / P_A CO_2 \tag{3}$$

V_A is the alveolar ventilation, 0.863 is a constant, VCO_2 is the CO_2 output and $P_A CO_2$ is the alveolar CO_2. The ability to maintain oxygen homeostasis is essential and the physiological systems compete against each other to provide enough tissue O_2 but also to maintain pH-homeostasis.

3.2 Cardiovascular system

To compensate for tissue hypoxaemia, the cardiovascular system must respond to maintain body functions. This is accomplished by increasing cardiac output, which is the product of stroke volume and heart rate [43]. Consequently, an increase in one of these variables will also lead to an increased volumetric flow rate. Upon ascent to hypoxic environments, the sympathetic nervous system activation leads to an initial increase in heart rate, cardiac output and blood pressure via the release of stress hormones [40, 44]. Stroke volume remains low in the first hours which is a consequence of reduced blood plasma volume because of bicarbonate diuresis. This occurs as a result of the fluid shift from the intravascular space and the suppression of aldosterone [40]. Interestingly, the sympathetic nervous system activation remains increased even if one is well acclimatized to altitude [45]. In contrast to sympathetic activation, cardiac output decrease once a certain level of hypoxia is reached after several days [46]. After a few days, e.g., muscle tissue adapts and extracts more O_2 from the circulating blood by increasing the arterial–venous oxygen difference. This reduces the demand for higher cardiac output. Reductions in stroke volume can be attributed due to decreased plasma volume as well as the above-mentioned increased pulmonary vascular resistance. From the systemic circulation perspective, the endothelial autocoids nitric oxide and prostaglandins have received more attention as they are potentially mediating hypoxic vasodilation in the vessels [47]. Hypoxic-induced vasodilation will therefor quickly increase the blood flow to O_2-deprived tissues. Low $P_a O_2$ levels will increase Ca^{2+} concentration inside the endothelial wall which might lead to increased synthetization of vasodilating endothelial factors [48]. The smooth muscle cells of the blood vessels also have K^+ ATP-channels, that are activated once the ATP/ADP quotient drops due to hypoxia. As a result of the increased conductivity of K^+, the cell membrane is hyperpolarized, followed by relaxation of the vascular muscle cells and vasodilation. This is especially well evoked in coronary and vertebral vessels [49].

P_AO_2 is, as mentioned earlier, at sea level around 100 mmHg and will decrease at altitude. At sea level, around 96% of haemoglobin is bound to O_2 which can be seen in **Figure 2**. The oxyhaemoglobin dissociation curve plays a crucial role in O_2 transport and demonstrates the interaction between the oxygen carrying capacity of haemoglobin and changes in partial pressure of oxygen [50]. When P_AO_2 drops to 50 mmHg at altitude, only about 80% of haemoglobin sites are bound to O_2. The sigmoidal shape of the curve minimizes an abrupt decline in oxygen-carrying capacity of the blood. Another crucial adaptive process is, that the dissociation curve will shift to the left [51]. This is mediated by respiratory alkalosis and therefore rise in blood pH. This left shift causes that at a P_AO_2 of 50 mmHg, instead of 80%, around 90% of haemoglobin is bound to O_2. As a result, more oxygen is bound on haemoglobin and more oxygen can be unloaded to the tissues [52].

3.3 Cerebral system

The brain consumes around 20% of the available oxygen at rest and is very sensitive to insufficient O_2 supply [53]. The ability to process large amounts of oxygen (over a relatively small tissue mass) is necessary to support the high rate of ATP production to maintain an electrically active for the continual transmission of neuronal signals [54]. From this perspective, it is clear that hypoxia can have negative effects on cognitive function [55]. From the literature, it is well known that various factors have an important influence on cognitive impairment during hypoxia, in case they occur. These include the grade of hypoxia (e.g. altitude height), ambient temperatures, performing exercise tasks, individual physiological responses and the influence of P_B [56].

One of the most sensitive regions of the central nervous system is the cerebral cortex. However, acute exposure to extreme hypoxia can also cause changes within wide regions of the brain. Subtle changes in the white and gray matter were already observed during ascending Mt. Everest and K2, reducing movement control and planning [57]. Motor speed and precision are also negatively affected in altitude compared to sea level performance [58, 59]. The complexity of central execution

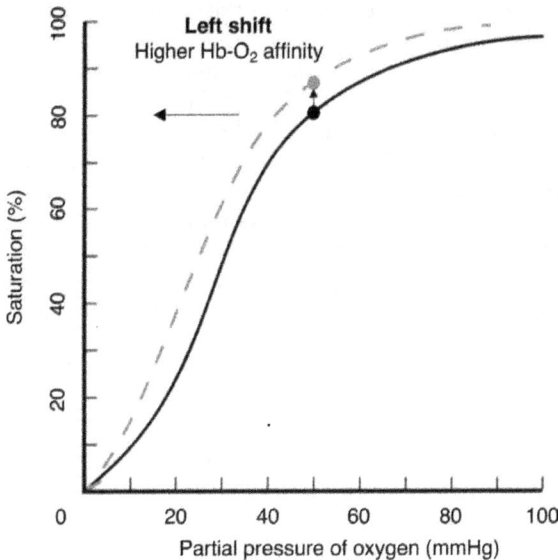

Figure 2.
S-shaped oxyhaemoglobin dissociation curve at sea level (solid black line). The curve is shifted left due to respiratory alkalosis under acute hypoxic exposure (broken gray line).

tasks seems to play an important role when cognitive impairment is evaluated. Cognitive impairment seems to be more prominent when complex tasks must be solved rather than simple tasks [60, 61]. Indeed, altitude accidents that occur under hypoxia might be more related to poor judgment of complex situations as a consequence of hypoxic depression of cerebral function. However, also small mistakes or even small increases in reaction time [62] can also have fatal consequences.

However, the underlying mechanisms, why cognitive performance can be impaired during hypoxia are not fully understood [61]. Cerebral circulation, which is the product of arterial oxygen content and cerebral blood flow, is dependent on the net balance between hypoxic vasodilation and hypocapnia-induced vasoconstriction. It is well documented, that cerebral blood flow is increased under acute hypoxia to maintain cerebral O_2-supply [54]. Cerebral blood flow increases, despite the hypocapnia, when arterial PO_2 is less than 60 mmHg (altitude greater than 2800 m). Although, interindividual varieties in cerebral blood are linked to individual variations in the ventilatory response to hypoxia [63], cerebral oxygen delivery and global cerebral metabolism are well maintained under moderate hypoxia. If cerebral oxygen consumption is constant, the question arises of what causes the cognitive impairment at altitude. Cognitive changes might be related to specific neurotransmitters that are affected by mild hypoxia (e.g., serotonin, dopamine). Furthermore, alterations in blood flow and sensory displeasure, hyperhomocysteinemia and potential neuronal damage, and a decrease in catecholamine availability combined with psychological factors appear to play a key role for reduction in cognitive function during hypoxia [61]. In case cerebral tissue oxygenation is not maintained, brain injury will occur with fatal consequences [35]. Compensatory hyperventilation, tachycardia and increased cerebral blood flow can partially maintain cerebral oxygen delivery, however, if these mechanisms work inadequately, the brain will be the first organ to be compromised.

4. Clinical consequences

This chapter aimed to give an overview of the main hypoxia types and the main physiological consequences. Hypoxia can occur due to occupational responsibilities, recreationally but also under pathological conditions. Ascend to altitude or exposure to environments that lower the P_iO_2 will have direct consequences to the entire body systems, however various modulators such as P_B, the severity of hypoxia, interindividual variability, health condition and others determine the physiological consequences and adaption processes. Exposing the body specifically to hypoxic environments can be used as a therapeutic tool, to increase sports performance or to achieve other goals [64]. However, it is important to precisely understand the different types of hypoxia and what consequences they have on the human body. Clinical manifestations of hypoxia underly inter-individual variations of cardiorespiratory and other physiological responses as well as the origin of hypoxia. In general, there are two major causes of hypoxia at the tissue level which are reduced blood flow to the tissues or reduced O_2 content in the blood itself [65, 66]. As a result, four main types of hypoxia arise. First, hypoxaemic hypoxia, where the O_2 transport to or through the alveoli is impaired [6]. Second, anemic hypoxia where the oxygen-carrying capacity is reduced due to e.g., severe blood loss, iron and folate deficiency, haemoglobin pathologies or functional insufficiency to carry O_2 [10, 12, 14]. Third, stagnant hypoxia where the transport of O_2 to the tissue is impaired while the blood may be sufficiently oxygenated [6]. Finally, histotoxic hypoxia exists, where the O_2 is delivered to the tissues but they are unable to utilize oxygen effectively [12].

It is important to understand, how these types influence oxygen delivery to the tissues. The product of O_2 content and blood flow is considered to reflect the oxygen delivery for the whole body (or to the individual organ system). As oxygen content is the sum of dissolved oxygen and that bound to haemoglobin, total oxygen delivery can be calculated according to Eq. (4):

$$DO_2 = (SaO_2 / 100 \times [Hb] \times 1.34 + P_aO_2 \times 0.023) \times \text{blood flow} \qquad (4)$$

DO_2 is the O_2 delivery (ml min^{-1}); P_aO_2 is the partial pressure of oxygen (kPa); S_aO_2 is the arterial oxygen saturation in percentage; Hb is the haemoglobin content (g dl^{-1}); 0.023 is the solubility of oxygen (in ml dl^{-1} kPa^{-1}); 1.34 is Hüfner's constant, the oxygen-carrying capacity of saturated haemoglobin (ml g^{-1}); and blood flow (i.e., cardiac output) in dl min^{-1} [67]. From this equation, it can be seen that hypoxaemic hypoxia (via reduced P_aO_2 and S_aO_2), stagnant hypoxia (via reduced blood flow) and anemic hypoxia (via reduced haemoglobin content) may cause tissue hypoxia, as these three types reduce oxygen delivery. In contrast, there is no oxygen delivery deficiency in histotoxic hypoxia but rather an impairment of the tissue to use O_2 [35]. Reduced oxygen tension, hypoventilation, ventilation-perfusion mismatch, right to left shunt and impaired diffusion of oxygen can all lead to hypoxia in the body [12].

The primary measurement to evaluate the hypoxic disease state is the analysis of arterial blood gas. Using this measurement, important parameters such as partial pressure of oxygen, partial pressure of carbon dioxide, acidity (pH), oxyhaemoglobin saturation and bicarbonate concentration in arterial blood can be assessed [68]. Management and treatment of persons under hypoxia should be started as soon as the evaluation has been successfully finished, and follows three categories: maintaining patent airways, increasing the oxygen content of the inspired air and improving the diffusion capacity [69–71]. Without adequate adaption processes and management, an imbalance between oxygen demand and oxygen delivery will occur leading to impaired homeostasis within the body. Therefore, healthcare practitioners (e.g., physiotherapists, sports scientists, exercise physiologists and others) should be able to understand the causes, types and consequences of hypoxia.

5. Conclusion

In this chapter, an overview is presented on the main types of hypoxia and the physiological consequences of the main systems. Hypoxaemic, anemic, stagnant and histotoxic hypoxia originate from different etiologies. Hypoxia to the tissues can be caused by any obstacle in the oxygen cascade, beginning from the O_2 molecule in the atmosphere, until being the final electron acceptor within the mitochondria to generate ATP. However, the adult compensatory mechanisms to counteract the acute hypoxic state are mainly based on our ability to hyperventilate, adequately adapt the cardiovascular response and to increase oxygen uptake to provide enough tissue O_2. This chapter might contribute to improving the understanding of the different types of hypoxia and to understand the physiological responses.

Conflict of interest

The author declares no conflict of interest.

Author details

Erich Hohenauer[1,2,3,4]

1 University of Applied Sciences and Arts of Southern Switzerland, Landquart, Switzerland

2 International University of Applied Sciences THIM, Landquart, Switzerland

3 University of Fribourg, Fribourg, Switzerland

4 Vrije Universiteit Brussel, Brussels, Belgium

*Address all correspondence to: erich.hohenauer@supsi.ch

IntechOpen

References

[1] Taylor CT. Mitochondria and cellular oxygen sensing in the HIF pathway. The Biochemical Journal. 2008;**409**(1):19-26

[2] Richalet J-P. CrossTalk opposing view: Barometric pressure, independent of , is not the forgotten parameter in altitude physiology and mountain medicine. The Journal of Physiology. 2020;**598**(5): 897-899

[3] Conkin J, Wessel JH 3rd. Critique of the equivalent air altitude model. Aviation, Space, and Environmental Medicine. 2008;**79**(10):975-982

[4] Roach RC, Hackett PH, Oelz O, Bartsch P, Luks AM, MacInnis MJ, et al. The 2018 Lake Louise Acute Mountain sickness score. High Altitude Medicine & Biology. 2018;**19**(1):4-6

[5] Wilson MH, Newman S, Imray CH. The cerebral effects of ascent to high altitudes. Lancet Neurology. 2009;**8**(2):175-191

[6] Pittman RN. Regulation of Tissue Oxygenation. Integrated Systems Physiology: From Molecule to Function to Disease. San Rafael (CA): Morgan & Claypool Life Sciences; 2011

[7] Hantzidiamantis PJ, Amaro E. Physiology, Alveolar to Arterial Oxygen Gradient. Treasure Island (FL): StatPearls; 2021

[8] Patel H, Kerndt CC, Bhardwaj A. Physiology, Respiratory Quotient. Treasure Island (FL): StatPearls; 2021

[9] Coleman MD. Chapter 2 - respiratory and pulmonary physiology. In: Duke J, editor. Anesthesia Secrets. Fourth ed. Philadelphia: Mosby; 2011. pp. 17-23

[10] Manninen PH, Unger ZM. Chapter 21 - hypoxia. In: Prabhakar H, editor. Complications in Neuroanesthesia. San Diego: Academic Press; 2016. pp. 169-180

[11] Brown LK. Hypoventilation syndromes. Clinics in Chest Medicine. 2010;**31**(2):249-270

[12] Bhutta BS, Alghoula F, Berim I. Hypoxia. Treasure Island (FL): StatPearls; 2021

[13] Rhodes CE, Varacallo M. Physiology, Oxygen Transport. Treasure Island (FL): StatPearls; 2021

[14] Lee EJ, Hung YC, Lee MY. Anemic hypoxia in moderate intracerebral hemorrhage: The alterations of cerebral hemodynamics and brain metabolism. Journal of the Neurological Sciences. 1999;**164**(2):117-123

[15] Phillips J, Henderson AC. Hemolytic Anemia: Evaluation and differential diagnosis. American Family Physician. 2018;**98**(6):354-361

[16] Sleijfer S, Lugtenburg PJ. Aplastic anaemia: A review. The Netherlands Journal of Medicine. 2003;**61**(5):157-163

[17] Chaudhry HS, Kasarla MR. Microcytic Hypochromic Anemia. Treasure Island (FL): StatPearls; 2021

[18] Nagao T, Hirokawa M. Diagnosis and treatment of macrocytic anemias in adults. Journal of General and Family Medicine. 2017;**18**(5):200-204

[19] Bleecker ML. Carbon monoxide intoxication. In: Handbook of Clinical Neurology. Vol. 131. 2015. pp. 191-203

[20] Ludlow JT, Wilkerson RG, Nappe TM. Methaemoglobinemia. Treasure Island (FL): StatPearls; 2021

[21] Benz EJ, Ebert BL. Chapter 43 - Hemoglobin variants associated with Hemolytic Anemia, altered oxygen affinity, and Methaemoglobinemias. In: Hoffman R, Benz EJ, Silberstein LE, Heslop HE, Weitz JI, Anastasi J, et al.,

editors. Hematology. Seventh ed. Philadelphia, PA, United States of America: Elsevier; 2018. pp. 608-615

[22] Cheung SS. Responses of the hands and feet to cold exposure. Temperature (Austin). 2015;**2**(1):105-120

[23] Feher J. 6.4 - oxygen and carbon dioxide transport. In: Feher J, editor. Quantitative Human Physiology. Second ed. Boston: Academic Press; 2017. pp. 656-664

[24] Wyman J. Facilitated diffusion and the possible role of myoglobin as a transport mechanism. The Journal of Biological Chemistry. 1966;**241**(1):115-121

[25] Graham J, Traylor J. Cyanide Toxicity. Treasure Island (FL): StatPearls; 2021

[26] Pauluhn J. Risk assessment in combustion toxicology: Should carbon dioxide be recognized as a modifier of toxicity or separate toxicological entity? Toxicology Letters. 2016;**262**:142-152

[27] Huzar TF, George T, Cross JM. Carbon monoxide and cyanide toxicity: Etiology, pathophysiology and treatment in inhalation injury. Expert Review of Respiratory Medicine. 2013;**7**(2):159-170

[28] Goldfarb-Rumyantzev AS, Alper SL. Short-term responses of the kidney to high altitude in mountain climbers. Nephrology Dialysis Transplantation. 2013;**29**(3):497-506

[29] Dietz TE, Hackett PH. 42 - high-altitude medicine. In: Keystone JS, Kozarsky PE, Connor BA, Nothdurft HD, Mendelson M, Leder K, editors. Travel Medicine. Fourth ed. London: Elsevier; 2019. pp. 387-400

[30] MacIntyre NR. Tissue hypoxia: Implications for the respiratory clinician. Respiratory Care. 2014;**59**(10):1590-1596

[31] Michiels C. Physiological and pathological responses to hypoxia. The American Journal of Pathology. 2004;**164**(6):1875-1882

[32] Duke HN, Killick EM. Pulmonary vasomotor responses of isolated perfused cat lungs to anoxia. The Journal of Physiology. 1952;**117**(3):303-316

[33] McMurtry IF, Davidson AB, Reeves JT, Grover RF. Inhibition of hypoxic pulmonary vasoconstriction by calcium antagonists in isolated rat lungs. Circulation Research. 1976;**38**(2):99-104

[34] Post JM, Hume JR, Archer SL, Weir EK. Direct role for potassium channel inhibition in hypoxic pulmonary vasoconstriction. The American Journal of Physiology. 1992;**262**(4 Pt 1):C882-C890

[35] Kane AD, Kothmann E, Giussani DA. Detection and response to acute systemic hypoxia. BJA Education. 2020;**20**(2):58-64

[36] Peers C, Kemp PJ. Acute oxygen sensing: Diverse but convergent mechanisms in airway and arterial chemoreceptors. Respiratory Research. 2001;**2**(3):145-149

[37] Lopez-Barneo J, Pardal R, Ortega-Saenz P. Cellular mechanism of oxygen sensing. Annual Review of Physiology. 2001;**63**:259-287

[38] Hallett S, Toro F, Ashurst JV. Physiology, Tidal Volume. Treasure Island (FL): StatPearls; 2021

[39] Paralikar SJ, Paralikar JH. High-altitude medicine. Indian Journal of Occupational and Environmental Medicin. 2010;**14**(1):6-12

[40] Hackett P, Roach R. High Altitude Medicine. Maryland Heights, MO, United States of America: Mosby; 2001. pp. 1-37

[41] Brinkman JE, Sharma S. Respiratory Alkalosis. Treasure Island (FL): StatPearls; 2021

[42] Lipsitz LA, Hashimoto F, Lubowsky LP, Mietus J, Moody GB, Appenzeller O, et al. Heart rate and respiratory rhythm dynamics on ascent to high altitude. British Heart Journal. 1995;74(4):390-396

[43] Talbot NP, Balanos GM, Dorrington KL, Robbins PA. Two temporal components within the human pulmonary vascular response to approximately 2 h of isocapnic hypoxia. Journal of Applied Physiology. 2005;98(3):1125-1139

[44] Mazzeo RS, Wolfel EE, Butterfield GE, Reeves JT. Sympathetic response during 21 days at high altitude (4,300 m) as determined by urinary and arterial catecholamines. Metabolism. 1994;43(10):1226-1232

[45] Domej W, Schwaberger G. Physiologie der mittleren, großen und extremen Höhen. In: Berghold F, Brugger H, Burtscher M, Domej W, Durrer B, Fischer R, et al., editors. Alpin- und Höhenmedizin. Berlin, Heidelberg: Springer Berlin Heidelberg; 2019. pp. 337-354

[46] Wagner PD. Reduced maximal cardiac output at altitude--mechanisms and significance. Respiration Physiology. 2000;120(1):1-11

[47] Busse R, Pohl U, Kellner C, Klemm U. Endothelial cells are involved in the vasodilatory response to hypoxia. Pflügers Archiv. 1983;397(1):78-80

[48] Franco-Obregon A, Lopez-Barneo J. Low PO2 inhibits calcium channel activity in arterial smooth muscle cells. The American Journal of Physiology. 1996;271(6 Pt 2):H2290-H2299

[49] Pohl U, de Wit C. Der Sauerstoff im Gewebe. In: Brandes R, Lang F, Schmidt RF, editors. Physiologie des Menschen: mit Pathophysiologie. Berlin, Heidelberg: Springer Berlin Heidelberg; 2019. pp. 365-375

[50] Brown JP, Grocott MP. Humans at altitude: Physiology and pathophysiology. Continuing Education in Anaesthesia, Critical Care and Pain. 2012;13(1):17-22

[51] Thorborg PAJ. CHAPTER 40 - blood gas analysis. In: Papadakos PJ, Lachmann B, Visser-Isles L, editors. Mechanical Ventilation. Philadelphia: W.B. Saunders; 2008. pp. 457-470

[52] Kenney WL, Wilmore JH, Costill DL. Physiology of Sport and Exercise. 6th ed. Champaign, IL, United States of America: Human Kinetics; 2015

[53] Attwell D, Laughlin SB. An energy budget for signaling in the grey matter of the brain. Journal of Cerebral Blood Flow and Metabolism. 2001;21(10):1133-1145

[54] Hoiland RL, Bain AR, Rieger MG, Bailey DM, Ainslie PN. Hypoxemia, oxygen content, and the regulation of cerebral blood flow. American Journal of Physiology Regulatory, Integrative and Comparative Physiology. 2016;310(5): R398-R413

[55] Nakata H, Miyamoto T, Ogoh S, Kakigi R, Shibasaki M. Effects of acute hypoxia on human cognitive processing: A study using ERPs and SEPs. Journal of Applied Physiology. 2017;123(5): 1246-1255

[56] Virués-Ortega J, Buela-Casal G, Garrido E, Alcázar B. Neuropsychological functioning associated with high-altitude exposure. Neuropsychology Review. 2004;14(4):197-224

[57] Di Paola M, Bozzali M, Fadda L, Musicco M, Sabatini U, Caltagirone C. Reduced oxygen due to high-altitude exposure relates to atrophy in motor-function brain areas. European Journal of Neurology. 2008;15(10):1050-1057

[58] Berry DT, McConnell JW, Phillips BA, Carswell CM, Lamb DG,

Prine BC. Isocapnic hypoxemia and neuropsychological functioning. Journal of Clinical and Experimental Neuropsychology. 1989;**11**(2):241-251

[59] Hornbein TF, Townes BD, Schoene RB, Sutton JR, Houston CS. The cost to the central nervous system of climbing to extremely high altitude. The New England Journal of Medicine. 1989;**321**(25):1714-1719

[60] Petrassi FA, Hodkinson PD, Walters PL, Gaydos SJ. Hypoxic hypoxia at moderate altitudes: Review of the state of the science. Aviation, Space, and Environmental Medicine. 2012;**83**(10): 975-984

[61] Taylor L, Watkins SL, Marshall H, Dascombe BJ, Foster J. The impact of different environmental conditions on cognitive function: A focused review. Frontiers in Physiology. 2015;**6**:372

[62] Cavaletti G, Tredici G. Long-lasting neuropsychological changes after a single high altitude climb. Acta Neurologica Scandinavica. 1993;**87**(2):103-105

[63] Ainslie PN, Poulin MJ. Ventilatory, cerebrovascular, and cardiovascular interactions in acute hypoxia: Regulation by carbon dioxide. Journal of Applied Physiology. 2004;**97**(1):149-159

[64] Heinonen IH, Boushel R, Kalliokoski KK. The circulatory and metabolic responses to hypoxia in humans - with special reference to adipose tissue physiology and obesity. Front Endocrinol (Lausanne). 2016;7:116

[65] Gaspar JM, Velloso LA. Hypoxia inducible factor as a central regulator of metabolism - implications for the development of obesity. Frontiers in Neuroscience. 2018;**12**:813

[66] Mesarwi OA, Loomba R, Malhotra A. Obstructive sleep Apnea, hypoxia, and nonalcoholic fatty liver disease. American Journal of Respiratory and Critical Care Medicine. 2019;**199**(7):830-841

[67] McLellan SA, Walsh TS. Oxygen delivery and haemoglobin. Continuing Education in Anaesthesia, Critical Care and Pain. 2004;**4**(4):123-126

[68] Sharma S, Hashmi MF. Partial Pressure of Oxygen. Treasure Island (FL): StatPearls; 2021

[69] Chen DW, Park R, Young S, Chalikonda D, Laothamatas K, Diemer G. Utilization of continuous cardiac monitoring on hospitalist-led teaching teams. Cureus. 2018;**10**(9):e3300

[70] Vali P, Underwood M, Lakshminrusimha S. Hemoglobin oxygen saturation targets in the neonatal intensive care unit: Is there a light at the end of the tunnel? (1). Canadian Journal of Physiology and Pharmacology. 2019;**97**(3):174-182

[71] Saito-Benz M, Sandle ME, Jackson PB, Berry MJ. Blood transfusion for anaemia of prematurity: Current practice in Australia and New Zealand. Journal of Paediatrics and Child Health. 2019;**55**(4):433-440

Chapter 4

Potential of Physical Activity-Based Intervention on Sleep in Children with and without Autism Spectrum Disorder

Thai Duy Nguyen

Abstract

Sleep problems are widespread, and sleep disorders are frequent in children with autism spectrum disorders (ASD). Physical activities (PA) are considered a practical, non-pharmacological approach for improving sleep. This study aims to explore the impact of PA on sleep in children with or without ASD. Seventy-five children were recruited, including 57 children with ASD and 18 typically developing (TD) children as control. Participants wore an accelerometer monitor (Sense Wear® Pro Armband 3, Body media) for 6 consecutive days and nights to assess sleep and PA. The results indicated ASD children had limited participation in PA compared with TD children (Total time for PA: 156 ± 79 vs. 216 ± 59 minutes on weekdays; 145 ± 93 vs. 178 ± 108 minutes on weekend). The children usually had more opportunities to participate in PA on weekdays and they tended to resist recommended bedtime (Sleep duration: 7.0 ± 0.8 vs. 9.6 ± 1.2 hours with ASD children; 7.1 ± 0.7 vs. 9.5 ± 1 hours with TD children). It also reported PA with moderate to vigorous intensity was better to improve sleep in children both with and without ASD. Finally, this study recommended promoting PA will help to improve sleep quality and reduce sedentary behaviors for children with ASD in particular and children in general.

Keywords: autism spectrum disorders, sleep disorder, physical activity, sleep quality

1. Introduction

Autism spectrum disorder (ASD) is known as a developmental disorder characterized by social communication difficulties and repetitive behavior. This is a complex syndrome related to genetic and environmental factors [1]. Great concerns about the high prevalence of poor sleep and the impact of sleep disturbance on ASD children are widely reported worldwide [2]. It is estimated that sleep disorders affect up to 80% of children with ASD compared with 10–25% of typically developing (TD) children [3, 4]. Children with ASD often face difficulties with sleep, and this has a strong relationship with daytime behavior problems; the most frequently reported issues include difficulty falling asleep, restless sleep, and frequent waking [5, 6]. Disturbed sleep could also exacerbate the core symptoms of ASD. Sleep education, environmental changes, behavioral interventions, and exogenous melatonin medication are frequently used for promoting and improving sleep quality [7, 8]. Improving the quality of sleep plays a critical role for children because sleep helps to optimize cognition, memory, behavioral adjustment, and learning [9].

Compared with TD children, the rate of sleep problems is higher in children with ASD. Significant impairments in social interaction and restricted behavior, combined with increased rates of motor problems, are frequently observed in individuals with ASD [10]. It may make these children less motivated and less likely to participate in physical activities (PA), leading to the risk of increased sedentary behavior (SB). It also may contribute to harmful health outcomes like overweight or obesity [11]. Recent reports showed difficulties in motor skills and motor capacities for ASD children compared with TD children, and limitations in PA may reduce opportunities for social interaction and learning in children with ASD [12, 13].

Physical activity is defined as any form of movement that leads to energy expenditure and is not performed in competition, including all daily activities, leisure activities, and exercises [14]. PA is indispensable for children's health. However, most children in the world do not participate in at least 60 minutes per day of moderate to vigorous physical activity (MVPA) as was recommended [15, 16]. Studies showed children diagnosed with ASD had PA levels lower than typically developed peers [17]. There are many individuals, social, and community barriers that make PA participation more difficult and may contribute to increased screen time by children with ASD [18]. Evidence has also been presented of PA decreasing negative behaviors and promoting positive behaviors. It improved social contacts and friendships and increased motor skills [11]. Thus, participation in PA is particularly essential for children with developmental disabilities, who could potentially benefit from increased PA and reduced SB; it has a positive impact on their development, quality of life, health, and future [9].

A reciprocal relationship between sleep and PA has been documented in children with and without ASD. Increasing exercise has been reported as helping produce better sleep quality, reduced weight, pain prevention, and improved mood in insomnia patients [19]. Adjusting factors of PA such as level, intensity, and duration of exercise has a positive effect on sleep quality [20]. Association between sleep patterns and PA levels suggests that being more physically active tends to support healthier sleep in children without disabilities [21]. Other studies also revealed a significant improvement in sleep efficiency, sleep onset latency, sleep duration, and wake after sleep onset with increased PA. It highlighted the role of PA in improving sleep quality among children with ASD [22]. Accordingly, regular moderate-intensity PA is recommended to treat and prevent sleep disorders without using medications [23]. In contrast, sleep disorders can lead to reduced cognitive performance and PA, while increasing the risk of injury during exercises. Getting insufficient sleep has been identified as a risk factor associated with public health problems such as obesity, depression, and limited PA [19]. Also, poor sleep was associated with higher rates of repetitive behavior and had a negative effect on challenging behaviors [24].

The mechanism of how PA affects sleep is not yet fully understood. Therefore, new studies must be carried out to understand the benefit of PA in the promotion of sleep and understand better the physiological responses to sleep loss. Within the scope of this study, we wanted to explore the relationship between PA intensity and sleep quality, its specific impact on improving sleep parameters. Thereby, it is possible to establish an optimal PA plan as a non-drug intervention to improve sleep quality as well as the quality of life in children with ASD.

2. Method

This study was approved by the local Ethics Committee of the Hospital (N°A00-865 40). It was conducted according to the Declaration of Helsinki and registered on the Clinicaltrials.gov registry (N°CT: 02830022).

2.1 Subjects

Each subject and their parents received both written and oral information, and those that agreed to participate signed a consent form. Seventy-five children were recruited to participate in the study, including 57 children with ASD and 18 typically developing children as a control group. All of them attended regular schools. Diagnosis of ASD was performed by experienced physicians and psychologists, according to the Diagnostic and Statistical Manual of Mental Disorders 5th edition criteria [25]. The subjects were also assessed with the Autism Diagnostic Observation Schedule (ADOS) [26]. Intelligence Quotient (IQ) was estimated using the Wechsler Intelligence Scale for Children, 4th edition [27]. IQ criterion for children is IQ > 70, excluding children with intellectual disabilities (IQ < 70). Following the ethical guidelines, IQ scores and ADOS results were not provided to researchers. However, score certification of IQ > 70 for all ASD children in this study was confirmed by a clinical psychologist experienced in diagnosing children with ASD and autism. Children with psychiatric disorders, comorbid medical conditions, contraindications for exercise, and those taking medication were excluded from the study [9].

2.2 Actigraphy

Participants wore the accelerometer monitor (Sensewear® Pro Armband 3, Body media) for 6 days and nights (5 weekdays and 1 weekend day). Participants and their parents completed daily diaries to distinguish periods in which participants did not wear the accelerometer (shower or bath, swimming, or other water activities). Time not wearing the device was excluded from the analysis. The monitoring device used in this study is a bi-axial accelerometer, worn on the right arm triceps. It can estimate energy expenditure based on algorithms from measured parameters such as acceleration, heat flux, galvanic skin response, skin temperature, near-body temperature, and demographic characteristics like sex, age, height, and weight [9].

This device can measure sleep parameters such as sleep time, sleep latency, wake up time, wake after sleep onset (WASO), and sleep efficiency. It was also used to calculate the parameters related to PA, such as the time spent for PA with different intensities and energy expenditure for PA. In the experiment period, children slept in their own bedroom, and their parents often described a consistent bedtime routine.

2.3 Child sleep diary

Children and their parents recorded information related to sleep each night for 6 consecutive days. It included bedtime (the time when the children went to bed each evening), wake up time (the time when the children woke up each morning), and sleep time (parents' estimation of duration of the children's sleep time each night).

Parents of the children also completed other questionnaires on sleep, physical activity, and parental assessments.

- The children's sleep habits questionnaire (CSHQ): This is the most used questionnaire to evaluate the sleep of ASD children, translated into French. CSHQ included 45 items with scores rated from 1 to 3, and was divided into eight subscales (bedtime resistance, sleep onset delay, sleep duration, sleep anxiety, night wakings, parasomnias, sleep-disordered breathing, and daytime sleepiness). CSHQ's total score was calculated and compared with the threshold value of 41. CSHQ's total score higher than 41 indicated sleep disorders or low sleep quality [28].

- The physical activity questionnaire for children (PAQ-C): PAQ-C provided general estimates of the level of PA and sedentary activities over the previous 7 days in children. It contained nine items with scores rated from 1 to 5. The final PAQ-C summary score was calculated to reflect PA levels, respectively Light (score = 1), Moderate (score = 2–4), and Vigorous (score = 5) [29].

- The questionnaire assessed parental awareness of children's sports practice. These questionnaire results helped to assess barriers to PA (time, economics, and emotion) and other values like the importance of sports practice, children's athletic level, parental support, and parents' sports practice.

2.4 Statistics

The data collected by actigraphy, sleep diary, and questionnaires were processed by specialized software (Sensewear Software) and Excel software in different ways. We calculated the average values of weekdays (WD), weekends (WE), and all days (AD) during the experiment for each child. Data used for statistical analysis were the mean values ± standard deviation. The differences of data between questionnaires and actigraphy method; WD and WE measured by actimetry in children with and without ASD were compared by R software (t-test, Pearson's chi-squared, one-way test with significance considered as $p < 0.05$). The relationship between factors related to sleep and PA was assessed by correlation and linear regression with significance considered as $p < 0.05$ (R software). Finally, principal component analysis (PCA) and agglomerative hierarchical cluster analysis (AHCA) were applied to classify individuals according to the component group (R software). In this study, we had $n = 75$ observations (57 children with ASD, 18 control children) and $p = 24$ predictors (anthropometric characteristics and data from monitoring devices). We used the elastic net method and appropriate criteria to select variables [12]. Finally, only 17 pertinent variables were used for PCA and AHCA.

The factors related to sleep used for analysis were total time on the bed (h), sleep duration (h), sleep quality index (%), bedtime resistance (min), sleep latency (min), wake-up time resistance (min), awakening latency (min), and wake after sleep onset (min). The factors related to PA used for analysis were sleep energy expenditure (kcal), total PA energy expenditure in 24 h (kcal), MVPA energy expenditure in 24 h (kcal), sedentary PA energy expenditure in 24 h (kcal), total time for PA (min), time for sedentary PA (min), time for moderate PA (min), time for MVPA (min), time for vigorous PA (min), time for strong vigorous PA (min), and daily step number.

3. Results

No difference between children with or without ASD was found with regard to demographic characteristics. However, we observed ASD children had a low PAQ-C score and a higher CSHQ score than control children as reported by their parents. Also, sleep duration as collected by questionnaires was more than by monitoring device by around 2.6 hours ($p < 0.001$) in ASD children and 2.4 hours ($p < 0.001$) in control children (**Table 1**).

		ASD (n = 57)	Control (n = 18)
Demographic	Age (years)	10.9 ± 2.9	10.1 ± 2.2
	Weight (kg)	39.2 ± 15.6	35.7 ± 13
	Height (cm)	146.5 ± 15.9	145.9 ± 15.6
	BMI (kg/m²)	17.6 ± 3.9	17.4 ± 3.5
	Gender (male, %)	87.7%	100%
Survey	PAQ-C (questionnaire)	2.5 ± 0.8**	3.2 ± 0.6
	CSHQ (questionnaire)	47.9 ± 6.8**	44.4 ± 4.5
	Sleep duration (questionnaire)	9.6 ± 1.2	9.5 ± 1
	Sleep duration (actigraphy)	7.0 ± 0.8*	7.1 ± 0.7*

		Weekdays		Weekend	
		ASD	Control	ASD	Control
Sleep variables	Total time on bed in 24 h (h)	9.7 ± 0.8	9.9 ± 0.8	9.8 ± 1.2	9.4 ± 2.7
	Sleep duration in 24 h (h)	7 ± 0.9	7.1 ± 0.8	7.1 ± 1.2	6.8 ± 2
	Sleep quality index in 24 h (%)	74.7 ± 6.9	74.4 ± 7.4	75.3 ± 11.2	70.9 ± 19.4
	Bedtime resistance in 24 h (min)	12.9 ± 8.2	10.5 ± 4.4	9.9 ± 9.3	11 ± 6.6
	Sleep latency in 24 h (min)	13.6 ± 9.9	15.9 ± 6.6	14 ± 18.1	13.1 ± 11.1
	Wake-up time resistance in 24 h (min)	1.5 ± 3.2	0.2 ± 0.3	1.3 ± 5	0.8 ± 3
	Awakening latency in 24 h (min)	15.5 ± 8.4	19.3 ± 6.7	16.6 ± 20.4	18.2 ± 13.2
	Wake after sleep onset in 24 h (min)	115.3 ± 41.4	126.7 ± 49.3	120 ± 61.4	115.6 ± 73.7
	Sleep energy expenditure in 24 h (kcal)	291 ± 91.7	282.3 ± 83.7	290.4 ± 105.2	259.8 ± 88.2

Physical activity variables	ASD (*n* = 57)		Control (*n* = 18)	
Total PA energy expenditure in 24 h (kcal)	1584 ± 548	1650 ± 595	1565 ± 596	1511 ± 699
Sedentary PA energy expenditure in 24 h (kcal)	1154 ± 431	1013.5 ± 370	1153 ± 438	1004 ± 487
MVPA energy expenditure in 24 h (kcal)	415 ± 231$	594 ± 272	391 ± 276	445 ± 309#
Total time for PA in 24 h (min)	156 ± 79$	216 ± 59	145 ± 93	178 ± 108
Time for sedentary PA in 24 h (min)	1284 ± 79$	1223 ± 61	1293 ± 91*	1182 ± 311
Time for moderate PA in 24 h (min)	136 ± 66	167 ± 47	125 ± 75	150 ± 80
Time for MVPA in 24 h (min)	154 ± 78$	209 ± 59	142 ± 88	170 ± 98
Time for vigorous PA in 24 h (min)	18 ± 20$$	42 ± 19	17 ± 23	19 ± 25#
Time for strong vigorous PA in 24 h (min)	1.8 ± 3$$	6.3 ± 5.9	3.3 ± 9.7	7.9 ± 21.6
Daily step number in 24 h	12,400 ± 4691$	15,739 ± 3287	12,080 ± 6180	12,065 ± 5336#

Values are mean ± SD. BMI (body mass index). * *$p < 0.001$ significantly different between sleep duration by questionnaire and actigraphy.*
** *$p < 0.05$ significantly different between ASD group and control group by questionnaire.*
# *$p < 0.05$ significantly different between weekdays and weekends in the control group.*
$ *$p < 0.001$ significantly different between ASD group and control group on weekdays.*
$$ *$p < 0.05$ significantly different between ASD group and control group on weekdays.*
* *$p < 0.05$ significantly different between ASD group and control group on weekend.*

Table 1.
Demographic data, and characteristics of sleep and PA.

No difference was found on PA factors between weekdays and weekends in ASD children, but there were differences in control children. They were more active in PA participation on weekdays than weekends (**Table 1**). On weekdays, ASD children had less energy expenditure for MVPA and time for PA (moderate, moderate to vigorous, vigorous, and strong vigorous) than control children. Meanwhile, they had more time for sedentary PA than control children on both weekdays and weekends. No difference in sleep factors was found between the two groups (**Table 1**).

Principal component analysis (PCA) and agglomerative hierarchical cluster analysis (AHCA) were performed with 17 pertinent variables. PCA results indicated five main component groups which helped explain 86.1% of the variance, with an eigenvalue ≥1. Two clusters were determined with blue space representing ASD children and yellow space representing control children (**Figure 1**). Two children with ASD (17, 23) and one control child (70) were classified outside of these clusters. In this graphic representation, a child on the same side as a given variable obtained a high score for this variable. A low value for this variable was attributed to a child on the opposite side. The distribution of children in both groups was not focused on their specific clusters. ASD children had discrete distribution (blue cluster), while control children were more concentrated (yellow cluster).

Child #17 was characterized by a total of daily steps twice the group average (25,620 vs. 12,354 steps/24 h) and had the highest time for vigorous PA (117.87 vs. 17.87 min/24 h), strong vigorous PA (16.17 vs. 2.07 min/24 h) compared with the group average. Child #23 was characterized by a total PA energy expenditure two times higher than the group average (3826 vs. 1581 kcal/24 h) and a sleep latency multiplied by 2.5 compared to the other children (35.4 vs. 13.6 min/24 h). Child #70 was characterized by a total PA energy expenditure twice the group average (3470 vs. 1642 kcal/24 h) and the highest sleep energy expenditure of all children in the group (500 vs. 280 kcal/24 h).

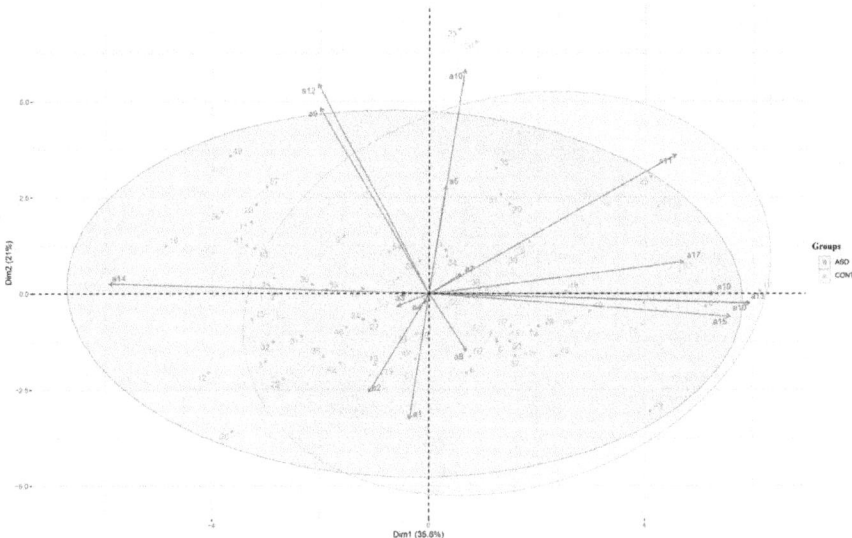

Figure 1.
Principal component analysis biplot of PA and sleep. ASD children are represented from 1 to 57, control children are represented from 58 to 75. (a1) Total time on bed (h); (a2) sleep duration; (a3) sleep quality index (%); (a4) bedtime resistance (min); (a5) sleep latency (min); (a7) awakening latency (min); (a8) wake after sleep onset (min); (a9) sleep energy expenditure (kcal); (a10) total PA energy expenditure (kcal); (a11) MVPA energy expenditure (kcal); (a12) sedentary PA energy expenditure (kcal); (a13) total time for physical activity; (a14) time for sedentary PA (min); (a15) time for moderate PA (min); (a16) time for MVPA (min); (a17) time for vigorous PA (min); and (a19) daily steps number.

(a)

(b)

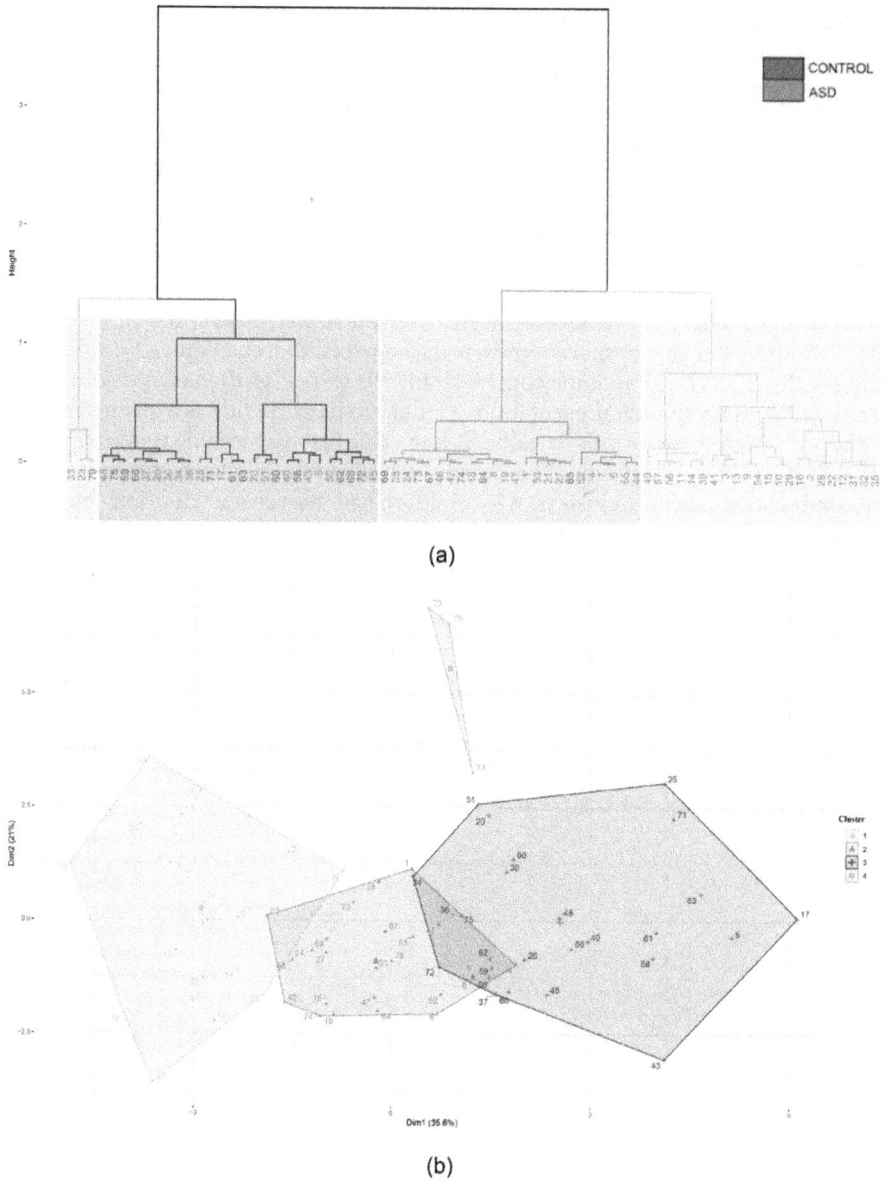

Figure 2.
(a) Dendrogram of individual's classification by AHCA. (b) Factor map of individual's classification by AHCA. Cluster 1 (saffron), cluster 2 (pink), cluster 3 (gray), and cluster 4 (green).

The results of an individual's classification by AHCA in **Figure 2** showed the ranking of each child and clusters in which they were classified. Based on the dendrogram graph, all children with similar characteristics in both the ASD group (green color) and the control group (red color) were classified into four different clusters. The characteristics of children in clusters were shown by comparing the result of the mean values of variables (**Table 2**).

- Cluster 1 (saffron): A total of 22 ASD children. This cluster is characterized by children with limited participation in PA and bad sleep. They had the lowest

			Cluster 1	Cluster 2	Cluster 3	Cluster 4
Sleep variable	Total time on bed in 24 h (h)	a1	9.88	9.65	9.96	7.92
	Sleep duration in 24 h (h)	a2	7.21	7.15	6.94	5.98
	Sleep quality index in 24 h (%)	a3	74.81	76.57	72.55	78.28
	Bedtime resistance in 24 h (min)	a4	11.15	14.30	10.07	15.55
	Sleep latency in 24 h (min)	a5	14.87	9.92	16.44	21.90
	Awakening latency in 24 h (min)	a7	17.08	13.30	19.18	14.85
	Wake after sleep onset in 24 h (min)	a8	116.29	110.30	134.80	64.19
Physical activity variable	Sleep energy expenditure in 24 h (kcal)	a9	337.99	246.42	266.88	447.91
	Total PA energy expenditure in 24 h (kcal)	a10	1594.62	1327.80	1653.54	3262.96
	MVPA energy expenditure in 24 h (kcal)	a11	247.81	364.77	637.31	1013.32
	Sedentary PA energy expenditure in 24 h (kcal)	a12	1343.80	945.53	976.69	2163.82
	Total time for physical activity in 24 h (min)	a13	79.44	162.02	247.64	182.54
	Time for sedentary PA in 24 h (min)	a14	1360.60	1277.52	1192.13	1257.52
	Time for moderate PA in 24 h (min)	a15	73.31	140.69	201.00	148.26
	Time for MVPA (min)	a16	79.04	159.19	241.40	178.99
	Time for vigorous PA in 24 h (min)	a17	5.73	18.50	40.39	30.73
	Daily steps number in 24 h	a19	8154.51	13,128.69	16,886.77	15,080.22

Values are mean of cluster.

Table 2.
Characteristics of ASD children in classification clusters.

PA (a11, a13, a15, a16, a17, and a19) except for sedentary PA (a12 and a14) and low sleep quality (a3) compared with other clusters.

- Cluster 2 (pink): A total of 24 members, including 18 ASD children and six control children. This cluster was characterized by children with moderate participation in PA and good sleep. They had high sleep quality (a3), high sleep duration (a2), and were moderate for PA compared with other clusters.

- Cluster 3 (gray): A total of 26 members, including 15 ASD children and 11 control children. This cluster was characterized by children with strong vigorous participation in PA and bad sleep. They had the highest PA (a13, a15, a16, a17, and a19) excepted for sedentary PA (a12 and a14) and the lowest sleep quality (a3) compared with other clusters.

- Cluster 4 (green): Total of three members, including two ASD children and only one control child. This cluster was characterized by children with moderate participation in PA and good sleep. They had the highest sleep quality (a3) and were vigorous for PA compared with other clusters.

4. Discussion

Most current studies still have not explained how PA affects sleep exactly and vice versa. PA and sleep influence each other through multiple complex interactions, both physiological and psychological [19]. PA is considered beneficial for improving sleep quality, but the effectiveness of PA-based interventions remains a question [30]. Conversely, sleep problems could increase symptoms of anxiety and depression, and thereby indirectly affect PA performance.

According to the survey results, we found a difference between children in the two groups. PAQ-C score indicated ASD children had PA levels lower than control children. CSHQ score presented higher sleep problems in ASD children than control children. Besides these, parent-reported estimated sleep time was higher than actigraphy-measured sleep time by 37.1% in ASD children and 33.8% in control children (**Table 1**). This demonstrated the children did not go to bed according to the schedule that their parents set. It was consistent with data collected from the monitoring device about total time in bed being more than sleep duration. Moreover, studies have reported on factors affecting sleep in a modern society like sleep environment, lifestyle habits, high-tech devices, physical activities, and learning activities [18, 31, 32]. Also, the downside of social development, children were free in their own rooms with little control from their parents. Therefore, they tended to resist recommended bedtime and fell asleep later. A typical example was when children went to bed, and even were lying in bed, but did not sleep, and instead, read stories or used smart entertainment devices.

Comparing results between children in the two groups, we observed ASD children had less participation in PA than control children. They had a low energy consumption for MVPA, daily step number, and total time for PA compared with control children on weekdays (**Table 1**). Also, time for MVPA, vigorous PA, and strong vigorous PA was equal to 74.7, 42.8, and 28.6% of control children on weekdays. Meanwhile, the time for sedentary PA was higher than in the control group both weekdays and weekends (**Table 1**). It proved ASD children often faced difficulties related to PA, especially MVPA, and tended to increase sedentary behaviors. Recent studies on the relevance of obesity and sedentary PA to sleep in adolescents and children with ASD showed similar results [33, 34]. No any significant differences in sleep between ASD children and control children, the reason was PA does not always affect sleep directly, as sleep also depends on control factors (such as age, health status, and mode and intensity of exercise intervention) or psychological factors [35, 36]. Our results also indicated control children had more active participation in PA (MVPA, vigorous PA, and daily steps) on weekdays compared to the weekend (**Table 1**). These differences may come from children usually going to school and participating in school activities on weekdays, while ASD children tended to be less sedentary and have less PA participation than TD children [37]. On the weekend, children were free to stay at home with their families, so they tended to have less PA participation.

PCA and AHCA analysis showed characteristics of PA and sleep in all children. We identified two relatively distinct clusters on the factors related to PA and sleep by PCA. One cluster included ASD children who had a positive correlation with sedentary PA, and the cluster with control children had a positive correlation with PA from moderate to vigorous, except for sedentary PA (**Figure 1**). Then, we presented the classification of individuals more clearly by AHCA, with four clusters determined to have a higher rate of ASD children than control children (**Figure 2**, **Table 2**). All of cluster 1 was ASD children; its characteristics were the lowest level of PA participation, highest sedentary behaviors, and bad sleep quality. This was consistent with the characteristics of children with ASD [6, 38]. Cluster 2 had 75% ASD children; its characteristics were a moderate level of PA participation, but they had better sleep in both duration and quality. Cluster 3 contained 57.7% ASD children; its characteristics were the highest level of PA participation, lowest sedentary behaviors, and worst sleep. Meanwhile, cluster 4 had 66.67% ASD children; its characteristics were a moderate level of PA participation and the best sleep quality. These classification results showed a difference between sedentary PA in ASD children and active PA participation in control children. Also, they indicated that PA with moderate to vigorous intensity was related to good sleep while limited participation in PA or strong vigorous PA was related to poor sleep. The positive effect of physical activity on sleep quality has also been discussed in studies of children with ASD [7, 39].

The findings in our study suggested the role of PA in improving sleep quality. Better sleep was the result of increased sleep duration and decreased sleep latency and wake after sleep onset. We should spend more daily energy consumption on MVPA, vigorous PA. It helps to increase PA and reduce SB. MVPA also was reported to enhance sleep quality by decreasing sleep latency and wake after sleep onset [40, 41]. Thus, we recommended increasing PA with moderate to vigorous intensity and sleep duration for improved sleep quality, especially with ASD children. This suggestion is in line with results reported in a comprehensive review of studies about the effects of PA on sleep quality with different PA intensities [23].

5. Conclusion

This study indicated ASD children tend to have low participation in PA and increased sedentary behaviors compared to control children. These children had more active PA participation on weekdays than the weekend, and they tended to resist bedtime by parents' request. Also, we provided evidence that PA with moderate to vigorous intensity can improve sleep quality, especially for children with ASD. It could be used as a non-pharmacological method to treat sleep disorders for ASD children. However, the nature and the magnitude of this impact are still controversial. Future studies need to clarify the mechanism of PA intensity effects on sleep quality. This way, they can give PA protocols based on reliable evidence to promote PA and prevent sedentary behaviors in children with ASD. This study also contributed to palliative treatment for children with ASD.

6. Limitations

The main limitation of our study was a disparity in the number of children between the ASD group and the control group. The sample sizes of the two groups were inconsistent. This affected the criteria on sleep quality used to distinguish children with and without ASD. These limitations need to be addressed in future studies.

Author details

Thai Duy Nguyen
NICVB, Ministry of Health of Vietnam, Hanoi, Vietnam

*Address all correspondence to: thainguyenduy@hotmail.com

IntechOpen

References

[1] Emberti Gialloreti L, Mazzone L, Benvenuto A, Fasano A, Garcia Alcon A, Kraneveld A, et al. Risk and protective environmental factors associated with autism spectrum disorder: Evidence-based principles and recommendations. Journal of Clinical Medicine. 2019;8:217

[2] Lord C. Taking sleep difficulties seriously in children with neurodevelopmental disorders and ASD. Pediatrics. 2019;143:e20182629

[3] Bangerter A, Chatterjee M, Manyakov NV, Ness S, Lewin D, Skalkin A, et al. Relationship between sleep and behavior in autism spectrum disorder: Exploring the impact of sleep variability. Frontiers in Neuroscience. 2020;14(211):1-13

[4] Moore M, Evans V, Hanvey G, Johnson C. Assessment of sleep in children with autism spectrum disorder. Children (Basel). 2017;4(72):1-17

[5] Deliens G, Peigneux P. Sleep–behaviour relationship in children with autism spectrum disorder: Methodological pitfalls and insights from cognition and sensory processing. Developmental Medicine and Child Neurology. 2019;61:1368-1376

[6] Souders MC, Zavodny S, Eriksen W, Sinko R, Connell J, Kerns C, et al. Sleep in children with autism spectrum disorder. Current Psychiatry Reports. 2017;19:34

[7] Andy C, Lee PH, Zhang J, Lai EW. Study protocol for a randomised controlled trial examining the association between physical activity and sleep quality in children with autism spectrum disorder based on the melatonin-mediated mechanism model. BMJ Open. 2018;8(e020944):1-7

[8] Richdale AL, Schreck KA. Examining sleep hygiene factors and sleep in young children with and without autism spectrum disorder. Research in Autism Spectrum Disorders. 2019;57:154-162

[9] Bricout VA, Pace M, Guinot M. Sleep and physical activity in children with autism spectrum: About 3 clinical cases. Austin Journal of Autism & Related Disabilities. 2018;4:1049

[10] Ohara R, Kanejima Y, Kitamura M, Izawa K. Association between social skills and motor skills in individuals with autism spectrum disorder: A systematic review. European Journal of Investigation in Health, Psychology and Education. 2020;10:276-296

[11] Jones RA, Downing K, Rinehart NJ, Barnett LM, May T, McGillivray JA, et al. Physical activity, sedentary behavior and their correlates in children with autism spectrum disorder: A systematic review. PLoS One. 2017;12:e0172482

[12] Bricout VA, Pace M, Dumortier L, Miganeh S, Mahistre Y, Guinot M. Motor capacities in boys with high functioning autism: Which evaluations to choose? Journal of Clinical Medicine. 2019;8(1521):1-15

[13] Odeh CE, Gladfelter AL, Stoesser C, Roth S. Comprehensive motor skills assessment in children with autism spectrum disorder yields global deficits. International Journal of Developmental Disabilities. 2020;89:1-11

[14] Thivel D, Tremblay A, Genin PM, Panahi S, Rivière D, Duclos M. Physical activity, inactivity, and sedentary behaviors: Definitions and implications in occupational health. Frontiers in Public Health. 2018;6(288):1-5

[15] Colley RC, Garriguet D, Janssen I, Craig CL, Clarke J, Tremblay MS. Physical activity of Canadian children and youth: Accelerometer results from

the 2007 to 2009 Canadian Health Measures Survey. Health Reports. 2011;**22**:15-23

[16] Parrish A-M, Tremblay MS, Carson S, Veldman SLC, Cliff D, Vella S, et al. Comparing and assessing physical activity guidelines for children and adolescents: A systematic literature review and analysis. International Journal of Behavioral Nutrition and Physical Activity. 2020;**17**:16-16

[17] Scharoun SM, Wright KT, Robertson-Wilson JE, Fletcher PC, Bryden PJ. Physical activity in individuals with autism spectrum disorders (ASD): A review. In: Fitzgerald M, Yip J, editors. Autism—Paradigms, Recent Research and Clinical Applications. London: IntechOpen; 2017. DOI: 10.5772/66680

[18] Must A, Phillips S, Curtin C, Bandini LG. Barriers to physical activity in children with autism spectrum disorders: Relationship to physical activity and screen time. Journal of Physical Activity and Health. 2015;**12**:529-534

[19] Chennaoui M, Arnal PJ, Sauvet F, Leger D. Sleep and exercise: A reciprocal issue? Sleep Medicine Reviews. 2015;**20**:59-72

[20] Healy S, Haegele JA, Grenier M, Garcia JM. Physical activity, screen-time behavior, and obesity among 13-year olds in Ireland with and without autism spectrum disorder. Journal of Autism and Developmental Disorders. 2017;**47**:49-57

[21] Benson S, Bender AM, Wickenheiser H, Naylor A, Clarke M, Samuels CH, et al. Differences in sleep patterns, sleepiness, and physical activity levels between young adults with autism spectrum disorder and typically developing controls. Developmental Neurorehabilitation. 2019;**22**:164-173

[22] Tse CYA, Lee HP, Chan KSK, Edgar VB, Wilkinson-Smith A, Lai WHE. Examining the impact of physical activity on sleep quality and executive functions in children with autism spectrum disorder: A randomized controlled trial. Autism. 2019;**23**:1699-1710

[23] Wang F, Boros S. The effect of physical activity on sleep quality: A systematic review. European Journal of Physiotherapy. 2019;**23**(1):1-8

[24] Abel EA, Schwichtenberg AJ, Brodhead MT, Christ SL. Sleep and challenging behaviors in the context of intensive behavioral intervention for children with autism. Journal of Autism and Developmental Disorders. 2018;**48**:3871-3884

[25] American Psychiatric Association. Diagnostic and Statistical Manual of Mental Disorders-V. Arlington, VA: American Psychiatric Publishing; 2013

[26] Lord C, Risi S, Lambrecht L, Cook EH Jr, Leventhal BL, Dilavore PC, et al. The autism diagnostic observation schedule-generic: A standard measure of social and communication deficits associated with the spectrum of autism. Journal of Autism and Developmental Disorders. 2000;**30**:205-223

[27] Wechsler D. Wechsler Intelligence Scale for Children. 4th Edition (WISC-IV). San Antonio, Tex: Harcourt Assessment; 2003

[28] Owens JA, Spirito A, McGuinn M. The children's sleep habits questionnaire (CSHQ): Psychometric properties of a survey instrument for school-aged children. Sleep (New York). 2000;**23**:1043-1052

[29] Kowalski KC, Crocker PR, Faulkner RA. Validation of the physical activity questionnaire for older children. Pediatric Exercise Science. 1997; **9**:174-186

[30] Rosenberger ME, Fulton JE, Buman MP, Troiano RP, Grandner MA, Buchner DM, et al. The 24-hour activity cycle: A new paradigm for physical activity. Medicine and Science in Sports and Exercise. 2019;**51**:454-464

[31] Reid Chassiakos Y, Radesky J, Christakis D, Moreno MA, Cross C. Children and adolescents and digital media. Pediatrics. 2016;**138**:e20162593

[32] Shochat T. Impact of lifestyle and technology developments on sleep. Nature and Science of Sleep. 2012;**4**:19-31

[33] Barnes DB. Short term examination of physical activity and sleep quality with children with autism spectrum disorder [Bachelor of Science (B.S.) Honors Undergraduate Theses]. University of Central Florida; 2019

[34] Mccoy SM, Morgan K. Obesity, physical activity, and sedentary behaviors in adolescents with autism spectrum disorder compared with typically developing peers. Autism. 2020;**24**:387-399

[35] Dolezal BA, Neufeld EV, Boland DM, Martin JL, Cooper CB. Interrelationship between sleep and exercise: A systematic review. Advances in Preventive Medicine. 2017;**2017**: 1364387

[36] Jung AR, Park JI, Kim H-S. Physical activity for prevention and management of sleep disturbances. Sleep Medicine Research. 2020;**11**:15-18

[37] Liang X, Li R, Wong SHS, Sum RKW, Sit CHP. Accelerometer-measured physical activity levels in children and adolescents with autism spectrum disorder: A systematic review. Preventive Medicine Reports. 2020; **19**:101147

[38] Bandini LG, Gleason J, Curtin C, Lividini K, Anderson SE, Cermak SA, et al. Comparison of physical activity between children with autism spectrum disorders and typically developing children. Autism. 2013;**17**:44-54

[39] Wachob D, Lorenzi DG. Brief report: Influence of physical activity on sleep quality in children with autism. Journal of Autism and Developmental Disorders. 2015;**45**:2641-2646

[40] Master L, Nye RT, Lee S, Nahmod NG, Mariani S, Hale L, et al. Bidirectional, daily temporal associations between sleep and physical activity in adolescents. Scientific Reports. 2019;**9**:7732

[41] Vanderlinden J, Boen F, Van Uffelen JGZ. Effects of physical activity programs on sleep outcomes in older adults: A systematic review. International Journal of Behavioral Nutrition and Physical Activity. 2020;**17**:11

Chapter 5

Exercise Mimetics: An Emerging and Controversial Topic in Sport and Exercise Physiology

Mohamed Magdy Aly Hassan ElMeligie

Abstract

Over the previous decade, there has been growing and fervent interest in scientific and commercial circles regarding the potential of bioactive compounds that mimic, or augment, the effects of exercise. These developments have given rise to the moniker 'exercise pills' or 'exercise mimetics'. The emergence of such orally-delivered bioactive compounds could hold substantial therapeutic value for combating metabolic disease. Such treatments might also present therapeutic value for morbidly obese individuals or those recovering from severe injury. This topic is not without controversy, however, as the search for a 'one size fits all' solution is not likely to bear fruit, given the complexity of the molecular and physiological mechanisms involved. The primary goal of this chapter is to explore the challenges of designing a pill that might reliably deliver the myriad and complex adaptations afforded by exercise training, with a focus on skeletal muscle. Furthermore, it will consider the issues, rationale, and practicality of implementing such therapeutics as a credible substitute to engaging in regular exercise training.

Keywords: exercise pill, bioactive, pharmaceuticals, human adaptation, skeletal muscle, homeostasis

1. Introduction

Physical exercise is recognized as a highly effective non-pharmaceutical intervention for a range of health conditions in humans. In the first instance, systematic review evidence (comprising millions of participants) has indicated that engagement in regular physical exercise is associated with a reduced risk for all-cause mortality, and in a dose-response manner [1]. Furthermore, it also has important benefits in the prevention and treatment of a range of chronic metabolic conditions [1], such as cardiovascular disease [2], diabetes [3], and cancer [4]. The benefits of regular physical exercise are not restricted solely to metabolic diseases, however. The whole-body homeostatic perturbations brought about by exercise-induced stress also encompass the central nervous system, skeletal muscle, skin, oxygen transport processes, and hepatic function [5]. An important observation is that the relationships between physical activity and health outcomes tend to be curvilinear, in that clinically relevant health benefits can be obtained from relatively little amounts of physical activity [1].

Despite its wide-ranging, multifaceted, and complex health benefits, almost one third of the global population over 15 years of age fails to meet the minimum prescription of physical exercise to obtain worthwhile health benefits [6]. In the United States, 8.3% (95% confidence interval: 6.4–10.2) of deaths have been attributed to inadequate levels of physical activity [7], a sobering statistic when considering the modifiable nature of this risk factor [8]. Yet more worrisome is the growing trend towards increasing sedentary behaviors (i.e., sitting time, computer use) over the previous decade [9, 10]; a fact made all the more severe by the ongoing COVID-19 pandemic and its associated government-mandated lockdown measures to protect public health [11]. Despite the seemingly global trend towards increased sedentariness and inadequate physical activity, impracticalities exist with regards to mandating an entire community, country, and/or global population to optimize their exercise habits [12]. It must also be noted that certain populations may not be able to engage in physical exercise due to injury, disease, or age-associated frailty, and thus would benefit from alternative solutions [13].

The potent effects of regular physical exercise on numerous important domains of human health have given rise to the notion of pharmacological compounds that mimic, or enhance, these effects. Such 'exercise mimetics' or 'exercise pills' have been touted as a potential, but not entirely probable, therapeutic solution [12, 14] for an otherwise challenging and ongoing public health problem. Although exercise brings about a range of physiological benefits to human health, compliance is often low and in certain groups may not be possible [15]. In recent decades, our understanding of the molecular determinants and physiological processes involved in exercise has improved at an alarming rate. This work has led to the emergence of chemical interventions that can induce the beneficial aspects of exercise, without necessitating actual skeletal muscle activity [15]. Such pharmacologic interventions may represent a viable strategy for addressing metabolic diseases associated with physical inactivity [16] or serve as an intermediary treatment for the morbidly obese or people recovering from serious injury [17]. The mechanistic basis for this supposition, and the opportunities and difficulties associated with such a strategy are the focal considerations of this chapter.

2. The identification of cellular and molecular targets in skeletal muscle

Skeletal muscle is the most abundant tissue in the human body, accounting for around 40% of total body weight, and is the most robustly activated organ in response to physical exercise [13]. In recent years, the effects of physical exercise on several molecular pathways and cellular targets in skeletal muscle have received significant attention. This investigative work has yielded numerous potential factors with relevance for 'exercise mimetic' applications in human health.

2.1 The AMPK-SIRT1-PGC1α pathway

The repeated muscular contractions brought about during physical exercise activate numerous signaling pathways in skeletal muscle, one of which is the AMPK-SIRT1-PGC1α axis that plays a key role in skeletal muscle energy metabolism and mitochondrial biogenesis [13].

2.1.1 AMPK

AMPK, or AMP-activated protein kinase, is a master regulator of energy homeostasis and metabolism within the cell. It is a heterotrimeric protein complex that comprises a catalytic subunit (α) and two regulatory subunits (β and γ) of which numerous isoforms exist [18]. AMPK integrates important signals from metabolic pathways and balances nutrient availability with energy demand. During exercise, muscle contractions deplete adenosine triphosphate (ATP), which reduces the ATP:AMP and ATP:ADP ratios within the cell, subsequently activating AMPK [19]. In skeletal muscle, the activation of AMPK induces a switch from anabolic cellular metabolism to a catabolic state of metabolism, blocking energy-consuming activities and promoting the synthesis of ATP from fatty acid oxidation, glycosylation, and glucose uptake [13]. These effects are mediated acutely by direct phosphorylation of metabolic targets, whereas a more chronic effect is brought about by gene transcription [13]. Inactivation of AMPK in skeletal muscle leads to the loss of oxidative fibers, suppressed fat metabolism, and impaired mitochondrial biogenesis [20].

Exercise is perhaps the most prominent physiological activator of AMPK in skeletal muscle. Acutely, exercise intensities above 60% of maximal aerobic capacity can induce AMPK activation, as can lower intensities of a prolonged duration [21]. Given its 'global' role as a regulator of cellular energy stress in response to environmental factors such as caloric restriction, physical exercise, and metabolic disease [22], AMPK has garnered substantial attention. It continues to represent a promising potential target for pharmaceutical intervention, particularly when considering its interactions with other effectors.

2.1.2 SIRT1

Sirtuin 1 (SIRT1) is a central regulator of metabolic processes in response to energy availability, and is primarily localized in the nucleus [23]. It is responsive to NAD^+ to NADH concentrations, and thus cellular energy availability, through its activation by AMPK [20], and it also senses changes in intracellular redox state [13, 23]. The activation of SIRT1 deacetylates and activates peroxisome proliferator-activated gamma coactivator 1-alpha (PGC1-α), upregulating its specific activity as a transcription factor on genes related to mitochondrial respiration and fatty acid metabolism [13, 24]. In conditions of overexpression or knock-out however, there is evidence to suggest that SIRT1 can also serve as a PGC1-α inhibitor, thus reducing mitochondrial activity [13]. In addition, during low nutrient availability, SIRT1 induces a shift in cellular metabolism towards fatty acid oxidation due to the scarcity of glucose [23]. SIRT1 helps to support cellular energy balance by inducing catabolic processes while inhibiting anabolic processes, thus maintaining energy homeostasis [23].

Physical exercise, specifically high-intensity interval training, has been shown to elevate SIRT1 activity in human skeletal muscle, and this was also associated with mitochondrial biogenesis [25]. Moreover, chronic exercise results in systemic adaptations that increase the levels of SIRT1 expression in the kidney, liver and brain in patients with neurodegenerative diseases, normalizing cellular processes and decreasing disease severity [26]. Defects in the pathways mediated in part by SIRT1 are known to lead to numerous metabolic disorders. Therefore, given the potential benefits of exercise-associated activation of SIRT1 for health and disease, the pharmacological manipulation of this target might elicit multiple benefits, and as such remains an area of focused attention.

2.1.3 PGC1-α

PGC1-α plays an integral role in cellular metabolism, serving as a co-activator of a vast range of downstream transcriptional factors and effectors involved in fatty acid oxidation and mitochondrial biogenesis [13]. In skeletal muscle, PGC1-α is activated by endurance exercise-mediated stimulation of p38 mitogen-activated protein kinase (MAPK) [13], subsequently enhancing mitochondrial biogenesis. Importantly, both acute and chronic physical exercise robustly increase the mRNA expression of PGC1-α in rodent muscle, therefore underscoring its importance in exercise training adaptations [20]. PGC1-α mediates the remodeling of skeletal muscle towards a more metabolically oxidative and less glycolytic fiber-type composition [27]. In muscle-specific PGC1-α knock-out models, impaired endurance, abnormal fiber composition, and inconsistent mitochondrial gene regulation have been documented [13], thus reinforcing the indispensable role of PGC1-α in exercise-mediated adaptations. It has also been posited that PGC1-α is a key factor in metabolic disorders, such as diabetes, obesity, and cardiomyopathy. These notions, allied to its regulatory action in lipid metabolism, make PGC1-α a potentially attractive target for pharmacological intervention [27].

2.2 PPARδ

Peroxisome proliferator-activated receptor delta (PPARδ) is a nuclear hormone receptor that transcriptionally regulates over 100 genes, playing a vital role in many biological processes [13], particularly those relating to energy balance [28] and fatty acid oxidation [29]. Although expressed abundantly in a range of metabolically active tissues, in skeletal muscle PPARδ is predominantly expressed in oxidative slow-twitch as opposed to glycolytic fast-twitch fibers. This expression is further induced by endurance-type exercise activity known to trigger an oxidative and/ or slow-twitch phenotype [20]. Its role in skeletal muscle includes the regulation of slow/fast-twitch fibers, lipid metabolism, oxidative processes, mitochondrial biogenesis, weight reduction, impairment of liver gluconeogenesis, and management of inflammatory processes [13, 20]. In rodent models, muscle-specific activation of PPARδ has demonstrated 'exercise-like' effects, such as increasing running endurance and guarding against diet-induced obesity and type II diabetes [30]. Furthermore, ablation of PPARδ in skeletal muscle induces an age-dependent loss of oxidative muscle fibers, running endurance, and insulin sensitivity [31], thus further reinforcing the role of PPARδ in fiber type remodeling. The weight of this evidence has led to the assumption that PPARδ is a central transcriptional regulator of oxidative metabolism the slow-twitch phenotype [20] thus representing a major 'exercise mimetic' target of interest.

2.3 ERRα/γ

Estrogen-related receptors (ERRs) are key nuclear regulators in mitochondrial energy metabolism [29], with their transcriptional activity determined by cofactors such as PGC-1α. ERRα is expressed in a range of tissues with high energy turnover, including skeletal muscle. ERRγ has a similar expression pattern but is selectively expressed in tissues with high rates of oxidation such as brain, heart, and muscle [29]. When PGC-1α is induced, ERRα plays a major role in controlling the mitochondrial biogenic gene network; in its absence, the ability of PGC-1α to enhance the expression of mitochondrial genes is drastically reduced [29].

In skeletal muscle, ERRα is expressed in oxidative and glycolytic fibers, whereas ERRγ is expressed in oxidative fibers only [29, 32]. Notably, ERRγ regulates oxidative metabolism not just in skeletal muscle, but in other tissues as well [33], and is a key determinant of the oxidative muscle fiber phenotype [15]. As such, it is highly expressed in type I skeletal muscle fibers. In rodent models, when ERRγ is transgenically expressed in type II fibers, it induces metabolic and vascular adaptations, in the absence of exercise [32]. These adaptations include prominent vascularization, the secretion of proangiogenic factors, and an alarming increase in endurance performance of 100% [32]. Given these characteristics, ERRγ is a prominent target for exercise mimetics because of its direct regulation of genes associated with mitochondrial oxidation, however there is a paucity of research on the topic [12]. When applied ectopically in glycolytic fibers, ERRγ instigates a shift in fiber type from glycolytic to oxidative, inducing mitochondrial biogenesis and bring about increased vascularization [29].

2.4 REV-ERBα

The nuclear receptor REV-ERBα (also known as nuclear receptor subfamily 1 group D member 1 (NR1D1), is highly conserved across species and plays important roles in circadian rhythm and metabolism [33]. In skeletal muscle, REV-ERBα are prominently involved in the regulation of mitochondrial biogenesis, mitophagy, the promotion of an oxidative fiber type, and the processes underpinning a higher endurance capacity [34]. In rodents, muscle-specific ablation of REV-ERBα was shown to blunt the AMPK-SIRT1-PGC-1α signaling pathway, decrease mitochondrial density, reduce oxidative phosphorylation activity, and downregulate genes associated with fatty acid metabolism [34]. Conversely, overexpression of REV-ERBα in C2C12 cells activated these regulators of training adaptations, increased mitochondrial biogenesis and induced fatty acid metabolism genes [35]. REV-ERBα also appears to play a role in modulating muscle mass, with its deficiency leading to increased expression of atrophy genes, and overexpression leading to diminished atrophy genes and increased fiber size [36]. Therefore, REV-ERBα has been identified as a promising pharmacological target for exercise mimetic applications.

3. 'Exercise mimetic' pharmacologic compounds as putative therapeutics for human health

The attractive properties of physical exercise for human health have garnered fervent interest from the pharmaceutical industry in recent years, likely due to the large and untapped market of sedentary individuals that, for varying reasons, do not engage in sufficient physical exercise [37]. Chiefly, the development of novel therapeutic approaches to replicate an exercise-training phenotype [38] by activating selected molecular targets—so-called 'exercise pills' [15] or 'exercise mimetics'—remains an area of substantial investment and effort. In using natural or synthetic compounds, it is possible to induce exercise-mimicking effects even in sedentary test animals [12], by activating molecular targets and genomic regulators such as those previously described. The foremost of these therapeutic approaches will now be discussed.

3.1 AMPK activators

AICAR, or 5-aminoimidazole-4-carboxamide-1-β-D-ribofuranoside, is at the forefront of several 'exercise-mimetic' compounds developed to target AMPK, a master regulator of cellular and organismal metabolism [39]. It is a well-known adenosine analog that is intracellularly converted into ZMP, which directly associates with and allosterically stimulates AMPK [12, 22] in a time- and dose-dependent manner [40]. Acutely, AICAR activates AMPK to bring about an increase in fatty acid oxidation, whereas chronic AICAR treatment promotes skeletal muscle fiber type transition from fast- to slow-twitch, and increases the expression of enzymes associated with aerobic respiration [20]. This fiber type reorganization, in concert with mitochondrial biogenesis, has been shown to significantly increase exercise performance (by an unexpected 44%) in sedentary mice following AICAR administration alone [39]. AICAR also induces skeletal muscle glucose uptake by effecting the translocation of the GLUT4 receptor to the sarcolemma [22]. These findings highlight the potential of AICAR as a potential agent to address the insulin resistance seen in type II diabetes.

Metformin is a drug of the biguanide class known to function in an AMPK-dependent manner, and is one of the most broadly available antidiabetic agents presently available [22]. It represents a first line medication used to treat type II diabetes, and activates AMPK in the liver through inhibition of mitochondrial complex I, which concomitantly reduces cellular ATP generation [12]. The glucose-lowering action of metformin is at least partly mediated by the activation of AMPK [38]. Although the mode of activation is different, metformin activates AMPK in a similar manner to AICAR, and they both have similar roles in hepatic glucose production [13]. In diabetic patients, metformin can reduce blood pressure and also improve multiple cardiovascular risk factors in obese individuals [13]. It may also possess anti-inflammatory properties, the specifics of which are still being explored.

3.2 Resveratrol

Resveratrol, a naturally occurring plant-derived polyphenol, is recognized as an activator of SIRT1 and AMPK [13, 22], but has multiple biological targets [20]. In yeast it has been shown to promote longevity, whereas in rodents this capacity is uncertain [33]. Although abundant in the human diet, resveratrol is perhaps most notably consumed in the seeds and skin of grapes [13]. It is highly lipophilic but has scarce bioavailability; nevertheless it is capable of extracellular, intracellular, and nuclear interactions [13]. Its role on the SIRT1-AMPK axis, as well as PGC1-α [38, 41], has received interest as a potential metabolism-regulating, 'exercise mimetic' compound. However, evidence in rodents is conflicting and it has been postulated that resveratrol might actually improve performance when used in synergy with exercise, rather than as a substitute [13]. In human clinical trials, resveratrol was shown to induce the expression of SIRT1 and AMPK in skeletal muscle, albeit in obese type II diabetic males [42]. From the perspective of exercise performance, resveratrol administration suppressed exercise-dependent improvements in aerobic respiration in aged inactive males, thus blunting the beneficial effects of training [43]. Therefore, further research is needed to cogently understand the mechanisms of action and optimal dose before it can be recommended in 'exercise mimetic' applications. It should also be noted that novel, more potent synthetic activators of SIRT1, such as SRT1720, have been developed that might represent promising candidates for application in a clinical setting [33], although research on these compounds is still in its infancy.

3.3 GW501516

The compound GW501516 is a selective agonist of PPARδ, and was initially developed for possible beneficial applications in metabolic and cardiovascular diseases [13]. However, pre-clinical work in animals highlighted its carcinogenic effects in multiple organs and the compound was subsequently abandoned [44]. Nevertheless, numerous studies from the past decade have linked this drug with potential 'exercise mimetic' effects [39]. For instance, a metabolomic study in mice showed that GW501516 treatment enhanced exhaustive running endurance in both trained and untrained animals, by increasing the specific consumption of fatty acids and sparing blood glucose [45]. The expression of genes regulated by PPARδ, including PGC1-α and pyruvate dehydrogenase kinase 4 (PDK4) were also significantly increased following treatment, as were other markers of fatty acid metabolism in skeletal muscle. Importantly, in untrained mice the administration of GW501516 alone was sufficient to increase running endurance, even following just 1 week of provision. Similar findings have been previously reported, albeit without any benefits to endurance capacity, demonstrating that GW501516 establishes an endurance gene signature, sharing 50% of the gene expression pattern with exercise [39]. Elsewhere, GW501516 administration improved endurance function in a mouse model of myocardial infarction when compared to placebo, and preserved oxidative capacity and fatty acid metabolism [46]. Collectively, these findings suggest that the activation of PPARδ at least partially mimics the effects of exercise.

3.4 GSK4716

GSK4716 is a synthetic ERRγ agonist that can activate the receptor with a similar potency to that of its ligand PGC-1α [15, 47]. It robustly activates genes involved in mitochondrial biogenesis, fatty acid oxidation, and the tricarboxylic acid cycle (TCA) when used to treat primary muscle cells [12], and promotes an endurance-trained phenotype in mice [32]. However, there is a discrepancy between acute and chronic activation of ERRγ in ligand-treated primary muscle cells and transgenic animals, respectively [33]. Although GSK4716 has been heralded as a candidate 'exercise pill' [15], the aforementioned 'exercise-mimicking' effects have not been established *in vivo* [12], and the compound is not yet approved for human use.

3.5 SR9009

The synthetic REV-ERBα agonist SR9009 was developed at the Scripps Research Institute in 2012 and has been identified as an 'exercise pill' of promise [15]. A single injection of SR9009 brought about 'exercise-like' effects in rodents, such as enhanced mitochondrial activity and the induction of genes associated with fatty acid metabolism [34]. After 12 days, energy consumption was enhanced without changing the respiratory exchange ratio, and after 30 days mouse running performance was significantly prolonged. Despite these positive findings, REV-ERB independent effects on cell proliferation, metabolism and gene expression have been found in a double-ablation model [48]. Therefore, positive outcomes with respect to the physiological and molecular effects of exercise should be interpreted with a degree of caution. More importantly, SR9009 has not been approved for human use at the time of writing, however tests have been devised against its surreptitious use [49].

4. The challenges and controversies of developing an 'exercise pill'

A pharmacological method of replicating the multifaceted and complex effects of physical exercise would no doubt be of value to populations that for whatever reason cannot engage in physical activity, such as people with disabilities, disease, frailty, or injury. For example, it might serve as an avenue towards reengaging with physical exercise after a severe injury, or a 'stepping stone' for individuals that are morbidly obese. However, there are several important considerations that need to be addressed.

4.1 Can physical exercise be realistically replaced?

There are inherent dangers in a 'reductionist' approach to exercise mimetics, as rodent knockout models have shown that no single 'exercise gene' or signaling pathway exists [37]. Even though PGC-1α has, for example, been described as the 'master regulator' of endurance exercise adaptations, evidence suggests that it may not be a prerequisite for exercise training-induced mitochondrial adaptations [37]. The biological responses to acute and chronic physical exercise in humans are characterized by a high degree of physiological redundancy at the molecular, cellular, organ-system, and whole-body levels [50]. Furthermore, the exercise-induced skeletal muscle phenotype is independent of a chosen few genes, proteins, and signaling pathways [51, 52]. Therefore, irrespective of the promising research findings discussed above, it is extremely unlikely that the emergence of a single pharmaceutical compound will be able to deliver the myriad and complex physiological, metabolic, and homeostatic disruptions brought about by exercise [5]. The multiplicity of responses, at a macro, 'system-wide' level [37], have been described as too diverse for a single pharmaceutical approach to address, and therefore a 'one size fits all' panacea is unlikely to come forward. It appears then that there is no true replacement for actual exercise, at least at present, due to the distinct and multifaceted metabolic responses that take place, especially in skeletal muscle. Despite these reservations, 'exercise mimetics' might represent an avenue to obtain at least some of the important benefits in those unable to achieve adequate amounts of physical exercise [12]. However, it could be argued that improving adherence to existing evidence-based exercise guidelines and pharmaceutical strategies (e.g., statins for cardiovascular disease) would be a more fruitful and productive objective for the promotion of human health.

4.2 Doping implications for elite athletes

From the perspective of performance sport, 'exercise mimetics' raise important and challenging questions. PPARδ agonists were added to the WADA Prohibited List that became effective in 2009, with AICAR also banned in the same year. In 2012, both GW501516 and AICAR were moved to class S4 (hormone and metabolic modulators) [53], and at the time of writing, this is still the case. Both of these compounds have received significant media attention over the last decade. For example, in 2012 members of the Spanish cycling team, including the team doctor, were arrested in connection with an international network supplying AICAR, due to its effectiveness on performance [12]. Despite this, it must be emphasized that AICAR is not approved for therapeutic use anywhere in the world, given its status as an experimental compound. In a separate instance, Russian race walker Elena Lashmanova tested positive for GW501516 in 2014 and was subsequently sanctioned. A very stable drug, GW501516 possesses a long half-life and is therefore easily detected in blood and urine samples [12], which poses major consequences for athletes seeking to obtain this compound for performance enhancement. By way of comparison, resveratrol is a natural, albeit weak, compound that has been shown to improve endurance performance in animals, yet it is not a prohibited substance. This is likely due to its low bioavailability and lack of consistently beneficial effects in humans [12]. Therefore, due care and attention must be observed when selecting compounds in pursuit of performance enhancement to ensure compliance with the WADA Prohibited List and mitigate the risk of compromising one's career.

4.3 Side effects of human metabolic modulators

The constant activation of metabolic pathways by pharmaceutical means, so-called 'metabolic overdrive', could have undesirable health effects [37, 54]. For example, the chronic activation of AMPK (i.e., via AICAR) and concomitant inhibition of the mechanistic target of rapamycin (mTOR; a central regulator of protein synthesis and anabolism) could bring about a state of chronic catabolism, or breakdown [37]. This problem would be exacerbated if multiple exercise mimetics or pills, targeting diverse pathways, were consumed. More specifically, and with relevance to the exercise mimetics discussed above, GW501516 demonstrated serious toxicity and multi-organ carcinogenicity in rodent studies, whereas human clinical trials reported no adverse effects, likely due to the short duration and low dose administered [53]. Even the naturally occurring compound resveratrol has been associated with side effects in humans, albeit to a lesser extent than the synthetic compounds previously discussed. In *in vitro* studies, the concentration-dependent cytotoxicity of resveratrol has been demonstrated, with high doses associated with deleterious effects [55]. Although safe and well-tolerated at doses of up to 5 g per day in humans, diarrhea has been documented as a frequent side effect at doses of 2000 mg [56], and there may be implications of high dose resveratrol supplementation for people with underlying health conditions [55].

5. Non-pharmacological 'exercise mimetic' alternatives

There do exist non-pharmacological alternatives to 'exercise pills' that can potentially be applied to mimic the characteristics of exercise training. For example, neuromuscular electrical stimulation (NMES) has been used to induce involuntary muscle contractions and support the maintenance of muscle mass in injured athletes [57]. This can potentially serve as a surrogate for physical activity, as it has been shown to stimulate muscle protein synthesis rates in older men, and can ameliorate the muscle atrophy associated with limb immobilization to a certain extent [57]. In contrast to the pharmacological methods described above, NMES can maintain muscle mass without safety concerns or appreciable side effects [58], thus representing a potential strategy for mimicking, at least in part, the metabolic effects of physical exercise. These findings may have the most utility in clinical populations observing periods of bed rest or immobilization, by reintroducing a degree of muscle contraction. This activity can enhance muscle protein synthesis in the fasted and fed states, which might support muscle health during short-term periods of disuse in a clinical setting [59].

Acute passive heating has demonstrated some exercise mimetic properties in humans, namely type II diabetics, when implemented in proximity to an oral glucose tolerance test (OGTT) [60]. One-hour of passive heating in water at 40°C either 30 min before or 30 min after commencing an OGTT increased extracellular heat shock protein 70 in the blood and increased heart rate and total energy expenditure (via increased fat oxidation) [60]. However, passive heating did not affect blood glucose concentrations or insulin sensitivity compared with a control group. In skeletal muscle, there is preliminary evidence that chronic passive heating can promote hypertrophy in animal and human models, alongside augmented voluntary and involuntary strength [61]. With further study, passive heating might be a worthwhile non-pharmacologic and exercise mimetic strategy for people that are unable to complete sufficient exercise.

6. Conclusions

Exercise mimetics remains an area of considerable effort and inquiry but is not without its challenges and controversies. Although early clinical research has identified numerous promising molecular targets for pharmaceutical intervention, there is a lack of human clinical data to support their implementation. This, allied to the multifaceted nature of the human physiological response to exercise, and the redundancy inherent in such a response, suggests that a 'one size fits all' approach will be unlikely to manifest. As such, efforts should be focused on increasing adherence to existing evidence-based exercise guidelines and pharmaceutical interventions for the promotion of human health. Notwithstanding, it is possible that multiple pharmaceutical approaches could emerge in the future that target specific molecular pathways for cumulative benefit. These strategies may offer substantial value for populations unable, or unwilling, to engage in actual physical exercise. Nonetheless, the implications of exercise pills for doping in elite sport, and the potential side effects associated with the administration of these compounds for human health, are areas of cautious consideration for the next decade and beyond.

Acknowledgements

The authorship criteria are listed in our Authorship Policy: https://www.intechopen.com/page/authorship-policy.

Conflict of interest

The author declares no conflict of interest.

Author details

Mohamed Magdy Aly Hassan ElMeligie
Faculty of Physical Therapy, Department of Basic Sciences, October 6 University, 6th of October City, Egypt

*Address all correspondence to: mohamed.magdy.pt@o6u.edu.eg

IntechOpen

References

[1] Warburton DER, Bredin SSD. Health benefits of physical activity: A systematic review of current systematic reviews. Current Opinion in Cardiology. 2017;**32**:541-556

[2] Tian D, Meng J. Exercise for prevention and relief of cardiovascular disease: Prognoses, mechanisms, and approaches. Oxidative Medicine and Cellular Longevity. 2019;**2019**:3756750. DOI: 10.1155/2019/3756750

[3] Colberg SR, Sigal RJ, Yardley JE, et al. Physical activity/exercise and diabetes: A position statement of the American Diabetes Association. Diabetes Care. 2016;**39**:2065-2079

[4] Fong DYT, Ho JWC, Hui BPH, et al. Physical activity for cancer survivors: Meta-analysis of randomised controlled trials. BMJ. 2012;**344**:17

[5] Hawley JA, Hargreaves M, Joyner MJ, et al. Integrative biology of exercise. Cell. 2014;**159**:738-749

[6] Hallal PC, Andersen LB, Bull FC, et al. Global physical activity levels: Surveillance progress, pitfalls, and prospects. Lancet. 2012;**380**:247-257

[7] Carlson SA, Adams EK, Yang Z, et al. Percentage of deaths associated with inadequate physical activity in the United States. Preventing Chronic Disease. 2018;**15**:E38

[8] Thornton JS, Frémont P, Khan K, et al. Physical activity prescription: A critical opportunity to address a modifiable risk factor for the prevention and management of chronic disease: A position statement by the Canadian Academy of Sport and Exercise Medicine. British Journal of Sports Medicine. 2016;**50**:1109-1114

[9] Yang L, Cao C, Kantor ED, et al. Trends in sedentary behavior among the US population, 2001-2016. JAMA. 2019;**321**:1587-1597

[10] Guthold R, Stevens GA, Riley LM, et al. Worldwide trends in insufficient physical activity from 2001 to 2016: A pooled analysis of 358 population-based surveys with 1·9 million participants. The Lancet Global Health. 2018; **6**:e1077-e1086

[11] Stockwell S, Trott M, Tully M, et al. Changes in physical activity and sedentary behaviours from before to during the COVID-19 pandemic lockdown: A systematic review. BMJ Open Sport & Exercise Medicine. 2021;**7**:e000960

[12] Fan W, Evans RM. Exercise mimetics: Impact on health and performance. Cell Metabolism. 2017;**25**:242-247

[13] Guerrieri D, Moon HY, van Praag H. Exercise in a pill: The latest on exercise-mimetics. Brain Plasticity. 2017;**2**:153

[14] Gubert C, Hannan AJ. Exercise mimetics: Harnessing the therapeutic effects of physical activity. Nature Reviews. Drug Discovery. 2021; **2021**:1-18

[15] Li S, Laher I. Exercise pills: At the starting line. Trends in Pharmacological Sciences. 2015;**36**:906-917

[16] Warden SJ, Fuchs RK. Are "exercise pills" the answer to the growing problem of physical inactivity? British Journal of Sports Medicine. 2008; **42**:862-863

[17] Hunter P. Exercise in a bottle: Elucidating how exercise conveys health benefits might lead to new therapeutic options for a range of diseases from cancer to metabolic syndrome. EMBO Reports. 2016;**17**:136

[18] Kjøbsted R, Hingst JR, Fentz J, et al. AMPK in skeletal muscle function and metabolism. The FASEB Journal. 2018;**32**:1741

[19] Gowans GJ, Hawley SA, Ross FA, et al. AMP is a true physiological regulator of AMP-activated protein kinase by both allosteric activation and enhancing net phosphorylation. Cell Metabolism. 2013;**18**:556-566

[20] Matsakas A, Narkar VA. Endurance exercise mimetics in skeletal muscle. Current Sports Medicine Reports. 2010;**9**:227-232

[21] Richter EA, Ruderman NB. AMPK and the biochemistry of exercise: Implications for human health and disease. The Biochemical Journal. 2009;**418**:261

[22] Wall CE, Yu RT, Atkins AR, et al. Nuclear receptors and AMPK: Can exercise mimetics cure diabetes? Journal of Molecular Endocrinology. 2016;**57**:R49-R58

[23] Nogueiras R, Habegger KM, Chaudhary N, et al. Sirtuin 1 and sirtuin 3: Physiological modulators of metabolism. Physiological Reviews. 2012;**92**:1479-1514

[24] Pardo PS, Boriek AM. The physiological roles of Sirt1 in skeletal muscle. Aging (Albany NY). 2011;**3**:430

[25] Gurd BJ, Perry CGR, Heigenhauser GJF, et al. High-intensity interval training increases SIRT1 activity in human skeletal muscle. Applied Physiology, Nutrition, and Metabolism. 2010;**35**:350-357

[26] Radak Z, Suzuki K, Posa A, et al. The systemic role of SIRT1 in exercise mediated adaptation. Redox Biology. 2020;**35**:101467

[27] Liang H, Ward WF. PGC-1α: A key regulator of energy metabolism. American Journal of Physiology. Advances in Physiology Education. 2006;**30**:145-151

[28] Liu Y, Colby JK, Zuo X, et al. The role of PPAR-δ in metabolism, inflammation, and cancer: Many characters of a critical transcription factor. International Journal of Molecular Sciences. 2018;**19**:3339. DOI: 10.3390/IJMS19113339

[29] Fan W, Evans R. PPARs and ERRs: Molecular mediators of mitochondrial metabolism. Current Opinion in Cell Biology. 2015;**33**:49-54

[30] Luquet S, Lopez-Soriano J, Holst D, et al. Peroxisome proliferator-activated receptor δ controls muscle development and oxidative capability. The FASEB Journal. 2003;**17**:2299-2301

[31] Schuler M, Ali F, Chambon C, et al. PGC1α expression is controlled in skeletal muscles by PPARβ, whose ablation results in fiber-type switching, obesity, and type 2 diabetes. Cell Metabolism. 2006;**4**:407-414

[32] Narkar VA, Fan W, Downes M, et al. Exercise and PGC-1α-independent synchronization of type i muscle metabolism and vasculature by ERRγ. Cell Metabolism. 2011;**13**:283-293

[33] Handschin C. Caloric restriction and exercise "mimetics": Ready for prime time? Pharmacological Research. 2016;**103**:158

[34] Woldt E, Sebti Y, Solt LA, et al. Rev-erb-α modulates skeletal muscle oxidative capacity by regulating mitochondrial biogenesis and autophagy. Nature Medicine. 2013; **19**:1039-1046

[35] Fan W, Atkins AR, Yu RT, et al. Road to exercise mimetics: Targeting nuclear receptors in skeletal muscle. Journal of Molecular Endocrinology. 2013; **51**:T87-T100

[36] Mayeuf-Louchart A, Thorel Q, Delhaye S, et al. Rev-erb-α regulates atrophy-related genes to control skeletal muscle mass. Scientific Reports. 2017;7:1-11

[37] Hawley JA, Joyner MJ, Green DJ. Mimicking exercise: What matters most and where to next? The Journal of Physiology. 2021;599:791-802

[38] Carey AL, Kingwell BA. Novel pharmacological approaches to combat obesity and insulin resistance: Targeting skeletal muscle with 'exercise mimetics'. Diabetologia. 2009;52:2015-2026

[39] Narkar VA, Downes M, Yu RT, et al. AMPK and PPARδ agonists are exercise mimetics. Cell. 2008;134:405-415

[40] Sullivan JE, Brocklehurst KJ, Marley AE, et al. Inhibition of lipolysis and lipogenesis in isolated rat adipocytes with AICAR, a cell-permeable activator of AMP-activated protein kinase. FEBS Letters. 1994;353:33-36

[41] Lagouge M, Argmann C, Gerhart-Hines Z, et al. Resveratrol improves mitochondrial function and protects against metabolic disease by activating SIRT1 and PGC-1α. Cell. 2006;127:1109-1122

[42] Goh KP, Lee HY, Lau DP, et al. Effects of resveratrol in patients with type 2 diabetes mellitus on skeletal muscle SIRT1 expression and energy expenditure. International Journal of Sport Nutrition and Exercise Metabolism. 2014;24:2-13

[43] Gliemann L, Schmidt JF, Olesen J, et al. Resveratrol blunts the positive effects of exercise training on cardiovascular health in aged men. The Journal of Physiology. 2013;591:5047-5059

[44] Sahebkar A, Chew GT, Watts GF. New peroxisome proliferator-activated receptor agonists: Potential treatments for atherogenic dyslipidemia and non-alcoholic fatty liver disease. Expert Opinion on Pharmacotherapy. 2014;15:493-503

[45] Chen W, Gao R, Xie X, et al. A metabolomic study of the PPARδ agonist GW501516 for enhancing running endurance in Kunming mice. Scientific Reports. 2015;5:1-13

[46] Zizola C, Kennel PJ, Akashi H, et al. Activation of PPARδ signaling improves skeletal muscle oxidative metabolism and endurance function in an animal model of ischemic left ventricular dysfunction. American Journal of Physiology. Heart and Circulatory Physiology. 2015;308:H1078

[47] Wang L, Zuercher WJ, Consler TG, et al. X-ray crystal structures of the estrogen-related receptor-γ ligand binding domain in three functional states reveal the molecular basis of small molecule regulation. The Journal of Biological Chemistry. 2006;281: 37773-37781

[48] Dierickx P, Emmett MJ, Jiang C, et al. SR9009 has REV-ERB-independent effects on cell proliferation and metabolism. Proceedings of the National Academy of Sciences. 2019; 116:12147-12152

[49] Geldof L, Deventer K, Roels K, et al. In vitro metabolic studies of REV-ERB agonists SR9009 and SR9011. International Journal of Molecular Sciences. 2016;17:1676

[50] Joyner MJ, Dempsey JA. Physiological redundancy and the integrative responses to exercise. Cold Spring Harbor Perspectives in Medicine. 2018;8:a029660. DOI: 10.1101/ CSHPERSPECT.A029660

[51] Qi Z, Zhai X, Ding S. How to explain exercise-induced phenotype from molecular data: Rethink and reconstruction based on AMPK and

mTOR signaling. Springerplus. 2013;**2**:1-10

[52] Kupr B, Schnyder S, Handschin C. Role of nuclear receptors in exercise-induced muscle adaptations. Cold Spring Harbor Perspectives in Medicine. 2017;**7**:029835. DOI: 10.1101/CSHPERSPECT.A029835

[53] Pokrywka A, Cholbinsk P, Kaliszewsk P, et al. Metabolic modulators of PPAR-delta. Journal of Physiology and Pharmacology. 2014;**65**:469-476

[54] Weihrauch M, Handschin C. Pharmacological targeting of exercise adaptations in skeletal muscle: Benefits and pitfalls. Biochemical Pharmacology. 2018;**147**:211-220

[55] Shaito A, Posadino AM, Younes N, et al. Potential adverse effects of resveratrol: A literature review. International Journal of Molecular Sciences. 2020;**21**:2084. DOI: 10.3390/ijms21062084

[56] Salehi B, Mishra AP, Nigam M, et al. Resveratrol: A double-edged sword in health benefits. Biomedicine. 2018;**6**:91. DOI: 10.3390/BIOMEDICINES6030091

[57] Wall BT, Morton JP, van Loon LJC. Strategies to maintain skeletal muscle mass in the injured athlete: Nutritional considerations and exercise mimetics. European Journal of Sport Science. 2015;**15**:53-62

[58] Dirks ML, Wall BT, Snijders T, et al. Neuromuscular electrical stimulation prevents muscle disuse atrophy during leg immobilization in humans. Acta Physiologica. 2014;**210**:628-641

[59] Dirks ML, Wall BT, Van Loon LJC. Interventional strategies to combat muscle disuse atrophy in humans: Focus on neuromuscular electrical stimulation and dietary protein. Journal of Applied Physiology. 2018;**125**:850-861

[60] James T, Corbett J, Cummings M, et al. Timing of acute passive heating on glucose tolerance and blood pressure in people with type 2 diabetes: A randomized, balanced crossover, control trial. Journal of Applied Physiology. 2021;**130**:1093-1105

[61] Rodrigues P, Trajano GS, Wharton L, et al. Effects of passive heating intervention on muscle hypertrophy and neuromuscular function: A preliminary systematic review with meta-analysis. Journal of Thermal Biology. 2020;**93**:102684

Section 2

Energy and Human Performance

Chapter 6

Methodological Procedures for Non-Linear Analyses of Physiological and Behavioural Data in Football

José E. Teixeira, Pedro Forte, Ricardo Ferraz, Luís Branquinho, António J. Silva, Tiago M. Barbosa and António M. Monteiro

Abstract

Complex and dynamic systems are characterised by emergent behaviour, self-similarity, self-organisation and a chaotic component. In team sports as football, complexity and non-linear dynamics includes understanding the mechanisms underlying human movement and collective behaviour. Linear systems approaches in this kind of sports may limit performance understanding due to the fact that small changes in the inputs may not represent proportional and quantifiable changes in the output. Thus, non-linear approaches have been applied to assess training and match outcomes in football. The increasing access to wearable and tracking technology provides large datasets, enabling the analyses of time-series related to different performance indicators such as physiological and positional parameters. However, it is important to frame the theoretical concepts, mathematical models and procedures to determine metrics with physiological and behavioural significance. Additionally, physiological and behavioural data should be considered to determine the complexity and non-linearity of the system in football. Thus, the current chapter summarises the main methodological procedures to extract positional data using non-linear analyses such as entropy scales, relative phase transforms, non-linear indexes, cross correlation, fractals and clustering methods.

Keywords: complex systems, positioning, physiology, performance analysis, soccer

1. Introduction

Applying non-linear theory and approaches has been a growing research interest in sports sciences fields such as performance analysis [1–4]. It is assumed that time-based and team sports display non-linear characteristics [5–8]. Football is deemed as a complex and dynamic system where players perform intermittent movements in time-space coordination [9–11]. The prominence of this research topic is due to several factors, amongst which the shift in the paradigm from linear to non-linear frameworks that has been applied to a wide variety of fields and settings, besides the ready access to technology (e.g., tracking systems) providing large datasets, time series outputs and new time-motion approaches [12–14].

Nonlinear theory and complex sciences are disruptive of linear frameworks [15, 16]. Linear systems assume a linearity on time-varying case, an input-output statistic and a linear state feedback [17]. Considering the linear system theory, an internal and external structure developing feedback control strategies for simultaneous stabilisation of the system [16]. Based on this, theoretical models quantify the relationship between human movement (input) and performance (output), considering the athlete as a linear system [18, 19]. Desynchronization between internal (such as heart rate measure, perceived exertion and biochemical procedures) and external components (i.e., movement speed, body impacts, metabolic power, accelerations and decelerations) may affect the performance [18, 20]. Small changes in the inputs determine proportional and measurable changes in the output, reporting linearity characteristics such as controllability, observability and canonical structure [21]. These assumptions determines approaches focused on the linearity of the system, reporting an fitness-fatigue binomial with a related dependence on dose-response relationships [22, 23]. However, the accuracy of these theoretical models has been challenged for being feeble and for the lack of individualised measurement [23]. Moreover, the ecological dynamisms of informational contexts, social relations and human movement variability are not considered in linear analyses [24–26]. Human movement and collective behaviour are not characterised by the linearity of the systems (as in team sports, like football) and the linear theory could be deemed as a reductionist approach to the problem [5, 6, 26]. Thus, the individual and collective performance has been reported using a complex and dynamic perspective [26–31]. Under these assumptions, biological systems are characterised by non-linearity, interaction-dominant dynamics, emergent behaviour, self-similarity, self-organisation and a chaotic component [32]. Literature reports several nomenclatures for the topic as complex adaptive systems [8, 33, 34], complex and dynamic systems [6, 26] or non-linear and dynamical self-organisation systems [27, 34, 35].

The ready access to cutting-edge technology was another reason for this field of research to increase. Such technology eventually became more affordable and user friendly. The use of tracking data started by assessing the individual players' movement, and later integrated spatial-temporal patterns based on Cartesian and Euclidean references [13, 36, 37]. Over the last two decades, positional data has been verified in football training and match-play to assess the complexity of the systems inherent to the individual movements and collective coordination [31, 38, 39]. Positional dataset can be applied to measure both physical and tactical measures [10, 40–45]. However, analysis do not always integrate different performance indicators [10, 46, 47]. Usually, studies focused only in a single performance dimension, however, football is a multifaceted sport with the physical, tactical, and technical factors amalgamating to influence performance with each factor not mutually exclusive of another [47]. Integrating performance metrics remain rarely described in current literature, concerning football environments [36]. That creates issues in the performance analysis, leading to the fact that the integrative approaches remain understudied. This research gap may be a very important topic to enhance knowledge about the theoretical concepts, mathematical models and methodological procedures of the non-linear approaches to integrate physiological and behavioural data in football.

2. Theoretical concepts of the non-linear approaches

2.1 Football as a complex and dynamic Cartesian coordinates

Football is an invasion game characterised as a complex and dynamical systems with a goal-oriented adaptation amongst teammate and opponents [9, 48].

Previously, to measure and tracking player's movements, mapping tactical actions and modelling collective behaviour were time-consuming processes [2, 4, 49]. Observational and notational analysis had scarce technological and procedural means to support the occurrence of the large number of physical, technical and tactical actions of the football game [49–51]. The wearable technologies as tracking systems allowed real-time access the players' position on the field during training and competition [31, 38, 39]. Positional dataset can be captured at different frequencies by using tracking systems as global positioning systems (GPS) tracking systems [52, 53], micro-electromechanical systems (MEMS) [36, 54], local radio-based local positioning systems (LPM) [55, 56], computerised-video systems [57, 58] and tracking system [59, 60]. This is largely due to the high cost associated with its use, which restricts its use almost exclusively to professional settings in male players [20, 61]. The validity and accuracy of these time-motion methodologies is well documented with an excellent reliability (coefficient of variation, CV: 1.02–1.04%) [52, 53]. However, the integration of the different devices still needs further studies [60, 62, 63]. Using this techniques to collect data, the players' movements are possible to be framed in a Cartesian referential (football field), represented by time series of Cartesian coordinates (xx, yy) [36]. Also, this approach allows spatiotemporal patterns to be assessed with a physical and tactical significance for coaches, performance analysts and researchers [30, 36]. Capturing collective and tactical performance also requires knowledge of the tactical-strategic variables that mediate intra- and inter-team behaviour [40, 64, 65]. According to Duarte et al. [11], inter-player and team coordination reports mutual influence of each player on the behaviour of dyadic systems shaped emergent performance outcomes. Sampaio and Vitor [64] applied positional data to calculate mean position in relation to the geometric centre of the team. Thus, longitudinal and lateral directions are established by the players' dyads and geometrical centre of the team (i.e., team centroid) [40, 45]. Movement patterns and inter-player coordination as a key issue in non-linear signal processing method [6, 64, 66]. Length, width and centroid distance as measures of team's tactical performance were also applied in youth football by Folgado et al. [66]. Synchronisation and synergy are theoretical terms recurrently used to assess intra-team and inter-team coordination and assess spatial-temporal displacement for goal orientation and team success [11, 65]. Typically, this special-temporal displacement has been assessed by the distance between player's dyads [42, 45]:

$$D\left(a_{x_{(t)}, y_{(t)}}, b_{x_{(t)}, y_{(t)}}\right) = \sqrt{\left(a_{x_{(t)}, y_{(t)}}\right)^2 + \left(a_{x_{(t)}, y_{(t)}}\right)^2} \tag{1}$$

where D is the distance, a is the player, x and y are the coordinates, t is the time, and b is the teammate or opponent. Team dispersion can be measured using effective playing position (EPS), surface area or spatial exploratory indexes using Euclidean (planar) coordinates [67–69]. For their measurement dyads players are achieved to measure the magnitude of the variability, predictability, stability and/or regularity in the distance between players [11, 40]. However, the individual preponderance within the synergies and synchronicities of the team remains somehow challenging to measure [70]. Additionally, ball possession is a critical issue in this positional data modelling given the importance of distinguishing phases of play [44, 64]. Due to this fact, different levels of analysis must be considered such as micro, meso and macro [13, 71, 72]. Moreover, positional and behavioural data cannot be interpreted separately, because football performance is a multifactorial phenomenon [10, 46, 47].

2.2 Current performance metrics to measure physiological demands under a linear framework

The training process requires a systematic physiological and biomechanical stimulus to ensure optimal adaptations and an adequate performance [19]. Several theoretical frameworks have been developed to assess the quantity and quality of the training and competition demands [18, 19]. These training load-based consider the linear system theory, likewise dose-response relationship and fitness-fatigue binomial. Fitness-fatigue model approach was originally proposed by Bannister [73]:

$$p(k) = p^* + c_1 \sum_{i=1}^{k-1} u(i) \exp^{\frac{-(k-1)}{\tau_1}} - \sum_{i=1}^{k-1} c_2(i) u(i) \exp^{\frac{-(k-i)}{\tau_2}} \qquad (2)$$

were $p(k)$ is the measured fitness (or performance) with gain term (k) and time constant (τ_1) [23]. Cumulative effect of training as a key guidance to the individual athlete's performance [20, 74]. Training load concept has been developed under the assumption that the athlete is a linear system [74, 75]. It is possible to breakdown training load into external and internal load. The external load describes the work rate (i.e., physical or biomechanical output), regardless of the biochemical and psychophysiological response [20, 76]. In training environments, it is mainstream to reports the work rate [49, 77], workload [78, 79] and training load [18, 19]. When load measures are also applied to analyse match performance, quite often is noted the physical performance [47], activity profile [80, 81] or match running performance [82–84]. Wherefore, integrating load measure with other performance factors is a current research-practice gap when determining metrics with physiological and behavioural significance [10].

2.3 Physiological and behaviour dataset in football environments: a integration approach using non-linear procedures

Football performance, a multifactorial phenomenon, dependant on a variety of factors such as environmental, contextual, physical, technical, tactical and psychophysiological [46, 47, 83]. These factor are not mutually exclusive of one another, what makes relevant an integrated approach to provide holistic insights about performance analysis [47, 83]. On regular basis, each of these factors are analysed in isolation without taking the others into account, leading to 1-dimensional insights [83]. Bradley and Ade [47] proposed a theoretical model emphasising on high-intensity running efforts during match-play advocating a contextualization of these running-based actions amongst technical and tactical activities [47]. This becomes of utmost relevance considering the football game [9, 65]. This is what mediates the players' decision making throughout the game according a team strategy previously defined [13]. Several authors have tried to establish ecological approaches to evaluate training and match outcomes, including non-linear approaches [11, 65]. Non-linear analyses were fundamentally performed on competitive game [41, 85] and limited training tasks as small and large-side games [42–45, 67, 69]. It is therefore important to understand mathematical models and methodological procedures underpinning non-linear analyses to assess their significance in applied research and applied settings, identify possible research gaps to be explored and, be aware of potential limitations and criticisms **(Figure 1)**.

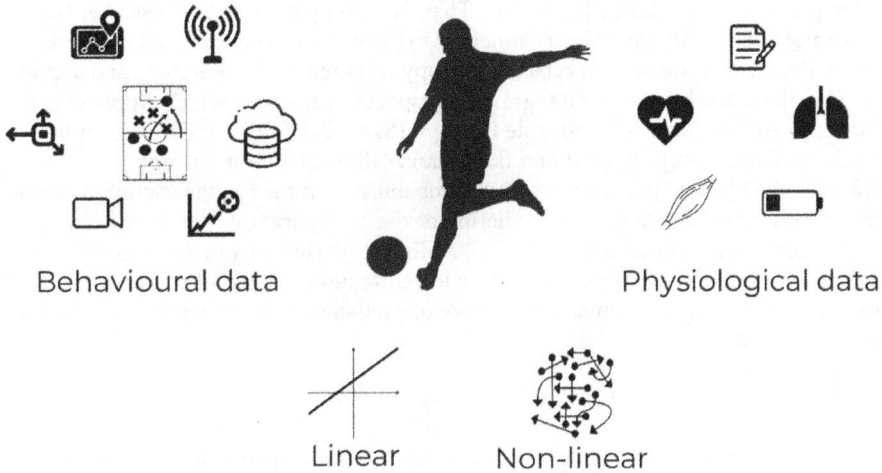

Figure 1.
Physiological and behavioural dataset in football environments.

3. Mathematical models and methodological procedures of the non-linear approaches

Non-linear approaches have been recurrently applied in football using complexity principles [14, 31, 36]. The informational context and spatiotemporal determinants that mediate players' perception and action are analysed by nonlinear and dynamic proprieties of the football game [6, 28]. It is assumed that environmental, task and organismic constraint influence individual and collective behaviour [5, 28]. This behaviour has a physiological cost over time that must be measured [18, 20, 76]. Informative content can be classified as different domains of variability, namely the frequency domain, the entropy domain and the scale-invariant domain [5, 86]. In biological systems, sequential time-series have become outstanding data analysis in multifaceted context [87–90]. According to Bravi et al. [86] the time-series data can be described through five different domains of variability: (1) statistical (i.e., statistical properties of the distribution in a stochastic process); (2) geometric (i.e., properties of the dataset shaped in a certain space); (3) energetic (i.e., energy or power of the time-series); (4) informational (i.e., degree of irregularity/complexity inherent to the order of the elements in a time-series); invariant (i.e., fractality or unchanging attributes over time or space). In football, time-series data application has been widely applied [14, 36]. Low et al. [36] organised the non-linear methods into measures of the regularity (or predictability) and synchronisation. Geometrical centre and periodic phase oscillators has been considered to analyse players' synchronisation and modelling the coordination of a team [14]. However, remains unclear the application of time-series data from an integrated approach perspective [10, 47, 83]. Thus, it is paramount to determine mathematical models and methodological procedures for non-linear time-series data analysis, bearing in mind an integrative approach. Therefore, the following subsections elaborate on the different non-linear mathematical models possible to apply in football.

3.1 Entropy

Entropy is a non-linear and informational parameter applied to describe variability, regularity or predictability of the movement/performance uncover the inter-player's interactions [86, 89, 91]. That, is entropy parameters describes the degree of irregularity/complexity inherent to the order of the elements in a time-series [86]. There are several types of entropy reported in the literature and applied in football research from the integrative perspective, amongst which Approximate Entropy (ApEn) [6, 11, 42], Sample Entropy (SampEn) [44, 92], Cross-sample Entropy (Cross-SampEn) [92] and Boltzmann-Gibbs-Shannon Entropy (ShannonEn) [43]. ApEn expresses the probability that the sequence configuration in a time-series data allows the prediction of the configuration from another sequence from a distance apart [89, 91]. ApEn was derivate from Kolmogorov-Sinai entropy and ranged amongst 0–2 where lower values correspond to more predictable and higher values stands more unpredicted patterns within time-series ($0 \le$ ApEn ≤ 2) [89].

$$\text{ApEn}(m, r, N) = \phi^m(r) - \phi^{m+1}(r) \tag{3}$$

where m is the window length distance amongst comparting time-series points, r is the similarity radius, N is the time-series length, and ϕ is the probability that points m distance within a tolerance level (r) [36, 90]. $C_i^m(r)$ measures how similar are the regularity of the data points in the window length (m) having regard to the following $\phi^m(r)$ [6, 36]:

$$\phi^m(r) = (N - m + 1)^{-1} \sum_{i=1}^{N-m+1} \ln C_i^m(r) \tag{4}$$

From a practical point of view, the imputed ApEn values should be computed with 2 to vector length (m) and 0.2 * standard deviation to the tolerance (r) [41, 45, 93]. ApEn reliability for short time series is low, providing relative consistency during changes in input parameters (m, r and N) [1, 89, 91]. For this reason, Richman and Moorman [89] developed SampEn where their logarithm is simpler with shorter time-series records than ApEn that is heavily dependent on the length record causing lacks of relative consistency. SampleEn logarithm has been was developed on the basis Grassberger et al. [94] reporting a larger window length and a greater relative consistency than ApEn [89–91]. Likewise, SampEn values close to zero indicates a regular or near-periodic time-series sequence, while the higher values reports a more unpredictable pattern ($0 \le$ SampEn $< \infty$) [88, 91, 92]. SampEn measures the negative logarithmic probability that two similar sequences of m points extracted within the tolerance limits (r) for a window length ($m + 1$) [36, 92].

$$\text{SampEn}(m, r, N) = -\ln \frac{\phi^{m+1}(r)}{\phi^m(r)} \tag{5}$$

Where, m, r, N, and ϕ retain the meaning from ApEn equation, whereby and $m + 1$ windows are compared for eliminating the self-matching bias in the ApEn [90]. $\phi^{m+1}(r)$ reports the total number of time-series sequences in a window length $m + 1$ as far the $\phi^m(r)$ expresses the total template in a length m within the aforementioned tolerance level (r) [36].

Multiscale entropy (MSE) as Cross-ApEn and Cross-SampEn was recently introduced from the primary entropy procedures (i.e. ApEn and Cross-ApEn) [90, 95]. Therefore, cross-entropy methods quantify the degree or complexity of coupling between two cross-sequences while the primary entropy techniques evaluated the asynchronism between two time series [87, 90]. Cross-SampEn remain a greater relative consistent than Cross-ApEn, being defined as long as one template finds a time-series sequence [89]. Mostly, Cross-SampEn has been recurrently used in football settings to measure players' synchrony [36, 92]. Cross-SampEn, the templates are chosen from the series u and compared with vectors from v, while the negative logarithm accounts the ratio of two average between outputs [36, 89]:

$$\text{Cross} - \text{SampEn}(m, r, N) = \left\{ -\ln \frac{\phi^{m+1}(r)}{\phi^m(r)} \right\} \tag{6}$$

Boltzmann-Gibbs-Shannon entropy was applied by Ric et al. [43] to measure temporal diversity and structural flexibility of the players. This entropy-based technique was originally applied by Balescu [96], reporting the configuration's probabilities as the large N in a relative frequency occurring in a stationary distributions described by Shannon [97]:

$$H(x) = \sum_{i=1}^{n} p_i \log_b p_i \tag{7}$$

where $p_i = n_i/N$, where n_i and N is the frequency and total number of the configuration, respectively [1, 43]. Predictable and unpredictable patterns were also reported as lower and higher entropy values, which is presented by absolute or normalised forma ($0 \leq H(x) \leq 1$) [36].

MSE techniques was applied in football by Canton et al. [44] to identify how positioning the goals in diagonal configurations on the pitch modifies the external training load and the tactical behaviour in youth football environments (i.e., small-sided games). The authors applied a SampleEn algorithm to compute entropy values in different timescales, calculating the area under for complexity index as reported in multiple entropy analysis for time-series [90, 98]. MSE techniques reports the point-to-point fluctuations over a time-series range [44, 90, 98] as:

$$y_j^{\tau} = \frac{1}{\tau} \sum_{i=(j-1)\tau+1}^{j\tau} x_i \tag{8}$$

Where, timescales is τ, y_i is the data point in the constructed time-series and x_i is a data point in the original time-series through a window length (N). Complex index (C_1) is calculated using the area under the constructed time-series and the original time scale curve [90, 98]:

$$C_1 = \sum_{i=1}^{N} \text{En}(i), 1 \leq y_j \leq \frac{N}{\tau} \tag{9}$$

Where, En is the reported entropy parameter at the time scale i and N is the total number of time scale used for the C_1 calculation [90].

3.2 Relative phase (Hilbert transform)

Relative phase was extensity reported in football within an integrative framework [40, 41, 67]. Using a Hilbert transform [$\phi(t)$], relative phase computes the difference between two signals reporting coupled oscillators and stable patterns of synchronisation, in-phase and anti-phase, when players moved in the same or opposite directions [36, 99]. Hilbert transform was originally introduce by Gabor

[100] with aim to measure the phase and amplitude for a signal. Palut and Zanone [99] configured a phase diagram plotting the imaginary part of the Hilbert transform:

$$\phi(t) = \phi_1(r) - \phi_2(r)$$

$$\phi(t) = \arctan \frac{H_1 s_2(t) - H_2(t) s_1(t)}{s_1(t) s_2(t) + H_1(t) H_2(t)} \tag{10}$$

Where, $H_1 s_2(t)$ and $H_2(t) s_1(t)$ were the Hilbert transform from the two compared signals [36, 99]. In football, longitudinal and lateral directions within the pitch were reported using near-in-phase synchronisation of each players' dyads [40, 41, 67]. The percentage of time spent in each near-in-phase mode of coordination was computed to verify in-phase synchronisation ($-30° \leq \phi \leq 30°$), anti-phase synchronisation ($-180° \leq \phi \leq -150°$ or $150° \leq \phi \leq 180°$) or without pattern synchronisation (all other ϕ degree) [36, 99]. Oscillation in football environment has been a recurrent non-linear approaches to process x- and y-directions and positions covered by football players in centroid and effective playing space zone [38, 101]. Sampaio and Vitor [64] displayed the relative phase post-test value to measure movement patterns and inter-player coordination. This study reported a higher regularity of the patterns with the increasing of the expertise level. There is a gap in understanding how the physiological dataset can influence the intra- and inter-team variability that needs to be further studied [47]. It remains to be understood how this varies across the different levels of the expertise.

3.3 Complex index

Non-linear parameters are often transformed into reliable complexes indices to measure complexity in football settings [43, 69, 93]. Dynamic overlap is a complex index used to compare time-series against the average cosine auto-similarity of the overlap between configurations within time lags [102]. It is an informational non-linear parameter that expresses how timescale statured in a dynamic behaviour using the exploratory breadth at different timescales [43, 69]:

$$<q_d(t)> = (1 - q_{stat})t^\alpha + q_{stat} \tag{11}$$

Where, $<q_d(t)>$ is the dynamic overload value, t is the time lag, q_{stat} is the horizontal asymptote and α expresses the gradient. Dynamic overlap tends to infinity wherefore predictable and unpredictable reflects zero an one values, respectively ($0 \leq q_d(t) \leq 1$) [36, 43, 69]. Additionally, trapping strength is the overall behavioural flexibility performed at lower and highest values of the time scale [43].

Another complex index reported in the literature is the stretch index, which can be defined as distances amongst players and the geometrical centre of the team [45, 85]. That is, this complex index measures the spatial expansion or contraction [103] as:

$$s_{ind} = \frac{\sum_{i=1}^{N} w_i d_i}{\sum_{i=1}^{N} w_i} \tag{12}$$

Here d_i is the distance between player i and their geometrical team centroid [103]. Stretch index can be expressed in meters, and is also computed by CV and entropy parameters [93, 104]. Otherwise, spatial exploration index (SEI) is obtained with the width and length displacements, computing the distance from each data point in the time-series according to geometrical centre [1, 2].

3.4 Correlation index

Windowed and cross correlation were also applied to assess collective behaviour through positional data in football training and match environments [36, 65, 92]. Cross correlation function is well-supported in the human movement research, wherein the overlapping time windows that enclosed the time-series sequence under analysis [105, 106]. Cross-Correlation function multiplied the point-to-point amongst two time-series data series, reporting the sum of the products and the respective relationship quantification [105]:

$$r_{xy} = \sum_{i=0}^{N-1} x_i y_i \qquad (13)$$

Where r_{xy} is the correlation across the window length of each analysed time-series (N); $x_i y_i$ are the data point of the calculated data series. Boker et al. [106] described cross-correlation with the pairwise dataset at two different time-series signals in accordance with:

$$r_{xy\tau} = \frac{1}{N-\tau} \sum_{i=1}^{N-\tau} \frac{(x_i - \bar{x})(y_{i+\tau} - \bar{y})}{\sigma(x)\sigma(y)} \qquad (14)$$

Where τ is the observations across cross-correlation (r) amongst time-series data point $(x_i y_i)$; $\bar{x}$ and $\bar{y}$ are the means, $\sigma(x)\sigma(y)$ are the standard deviations in the studied window length (N) reporting positives and negatives correlation $(-1 \leq r \leq 1)$ [105–107]. Pearson coefficient (r) was expressed as [105]:

$$r_{xy} = \frac{N(\sum xy) - (\sum x)(\sum y)}{\sqrt{[n \sum x^2 - (\sum x^2)][n \sum y^2 - (\sum y^2)]}} \qquad (15)$$

where, r_{xy} maintains the meaning to the previous equation. Person-moment correlation was used to estimate in-phase synchronisation (i.e., r values close to -1), anti-phase synchronisation (i.e., r values close to 1) and without pattern (i.e., $r = 0$) [36, 92]. It remains to be seen whether matrix correlation may be applied to correlate players synchronisation with physiological insight, since it was only applied to assess dynamic collective behaviours in isolation individual physiological demands [65, 92, 107]. Complementarily, a study associated the cross-correlation with the an vector coding technique to analyse coordination patterns between teams during offensive sequences that ended in shots on goal or defensive tackles [108]. The authors based their non-linear analysis the relative motion plot proposed by Sparrow et al. [109]:

$$\theta(i) = \arctan \left| \frac{\theta_2(i+1) - \theta_2(i)}{\theta_1(i+1) - \theta_1(i)} \right|, i = 1, 2, \dots, n-1 \qquad (16)$$

where i is the time-series data point in a right horizontal and n is the total frame for each timescale [108]. As well, the $\theta(i)$ is the coupling angles between in the second, third and fourth adjusted quadrant (i.e., $\pi - \theta_i$, $\pi + \theta_i$, $2\pi - \theta_i$) [36, 109]. Near-in-phase synchronisation were computed using this non-linear technique, expressing the in-phase ($22.5° \leq \theta_i < 67.5°$ or $202.5° \leq \theta_i < 247.5°$) and anti-phase ($22.5° \leq \theta_i < 67.5°$ or $202.5° \leq \theta_i < 247.5°$). Even more, attacking and defensive team phases are also reported by Moura et al. [108]. By distinguishing the phases of the play, this non-linear method could allow to understand how physiological demands can affect the intra- and inter-synchronisation in offensive and defensive phases.

3.5 Fractality

Fractal dimension is an invariant non-linear parameter characterised by the unchanged proprieties of the system over time and/or space [86]. Multifractal time series expresses different local scaling exponents for a time-series dataset scaling different exponents at different times [86, 110]. Few studies applied fractal dimension to predict stability and predictability of the football players along specific training tasks [110, 111] and competitive matches [110–112]. Fractal calculus (FC) was reported using Shanon and Grünwald-Letnikov approaches [111, 112]. Grünwald–Letnikov fractional differential consideres the matrix containing the multi-player positions [111, 113]:

$$X_\delta[t] = \begin{bmatrix} x_1[t] \\ \vdots \\ x_{N_\delta[t]} \end{bmatrix}, \, (x_i[t] \in R^2) \tag{17}$$

where N_δ expresses the number of players in the team N_δ across a time-series analysis. The matrix $X_\delta[t]$ is the planar positioning matric of player (i) in a target team (δ) at the a concrete time (t) [111, 113]. Shannon information was expressed by [112]:

$$I[P(x)] = -\log P(x) \tag{18}$$

whereby $I[P(x)]$ is the function between the cases $D^{-1}I[P(x)] = P(x)[1 - \log P(x)]$ and $D^1I[P(x)] = P(x)\left[1 - \frac{1}{P(x)}\right]$. I and D are the integral and descriptive operations [112].

A study applied multifractal dimension in football movement behaviour using Hausdorff dimension (D) [110, 112, 114]:

$$D(E) = \lim_{\varepsilon \to 0} \sup \frac{\log N_E(\varepsilon)}{-\log \varepsilon} \tag{19}$$

where N_E is the minimal diameter at the most ε needed to cover [114]. Fractional dynamics may tracking football players trajectory, which can be useful to increase the autonomy of tracking systems [113]. Additionally, fractal and multifractal analysis can be used to analyse the regularity and synchronisation of the team, as well the players' movement dynamics [110, 111, 113]. However, the use of fractal variables within an integrative approach remains little explored whereas it is important understanding the links between collective behaviour with the fractional properties of movement and its physical and physiological repercussions [47, 110]. Fractal proprieties may also increase the autonomy of the tracking systems to collect information in tracking systems such as making decisions based on it [111].

3.6 Clustering methods

Clustering methods have become popular in data mining in several research areas, including sports sciences [70, 92, 115]. Rokach and Maimon [115] was described the clustering methods in different typologies as hierarchical, partitioning, density-based, model-based, grid-based, and soft-computing methods.

Duarte et al. [92] pioneered applied a clustering method to measure overall and player team collective synchronisation in football. This method is derived from Hibert transform to calculate individual phase time-series and subsequently the cluster phase of these time-series by the natural exponential function [36, 92]. Originally, cluster phase analysis was proposed by Frank and Richardson [116] using Kuramoto's parameters for group synchronisation [117]. This clustering method calculates the mean and continuous group synchronisation $\left(\rho_{\text{group}} \text{ or } \rho_{\text{group},i}\right)$ whence individual's relative phase (ϕ_k) was measured in relation to the group measure [116]. After the Hilbert transform calculation, the continuous degree of overall team synchronisation was clustered [92]:

$$\rho_{\text{group}}(t_i) = \left|\frac{1}{n}\sum_{k=1}^{n} \exp\{i(\phi_k(t_i) - \overline{\phi_k})\}\right| \tag{20}$$

where overall team synchronisation $\rho_{\text{group}}(t_i) \in [0, 1]$ and mean degree to group synchronisation at every point-to-point of the time-series data (t_i) [36, 92]. $\rho_{\text{group},i}$ was computed by the inverse of the circular variance of relative phase of the cluster amplitude, $\phi_k(t_i)$, while $i = \sqrt{-1}$ as [92]:

$$\rho_{\text{group}}(t_i) = \frac{1}{N}\sum_{i=1}^{n} \rho_{\text{group},i} \tag{21}$$

Synchronisation cut-off values is zero to one representing synchronisation and unsynchronisation ($0 \leq \rho_{\text{group}} \text{ or } \rho_{\text{group},i} \leq 1$), respectively [36, 92]. Cluster phase analysis has not yet been applied to integrate physiological and behavioural data in football [38, 101].

Furthermore, average mutual information (AMI) was also applied to measure complexity of the football patterns and expresses the amount of information one random variable contains another [118]. AMI is calculated by relative entropy between probabilities distribution and the product midst two selected variables [36]. The mathematical equation described by Cover and Thomas [118] for the calculation of mutual information is:

$$I(x,y) = \sum_{X,Y \in A} P(x,y) \log\left(\frac{P(x,y)}{P(x)P(y)}\right) \tag{22}$$

Where (x) and (y) is the team's centroid movement coordinates and A is the space discretisation [119]. $P(x)$ and $P(y)$ are the marginality of the probabilities distributions [36]. The AMI identify the relationships between time-series points that are not detected by linear correlation [118, 119]. Similar correlation cut-offs values were applied to predict uncertainty in less values and independence in higher values ($0 \leq \text{AMI} < 1$) [119].

3.7 Frequency domain

Non-linear techniques as entropy measures can also be expressed in the frequency domain [86]. Several studies have evaluated the variability of movement comparing informational and frequency domains [10, 41, 42]. CV expresses the magnitude of the variability in the distance amongst players', expressed as percentage (%) [10, 93]:

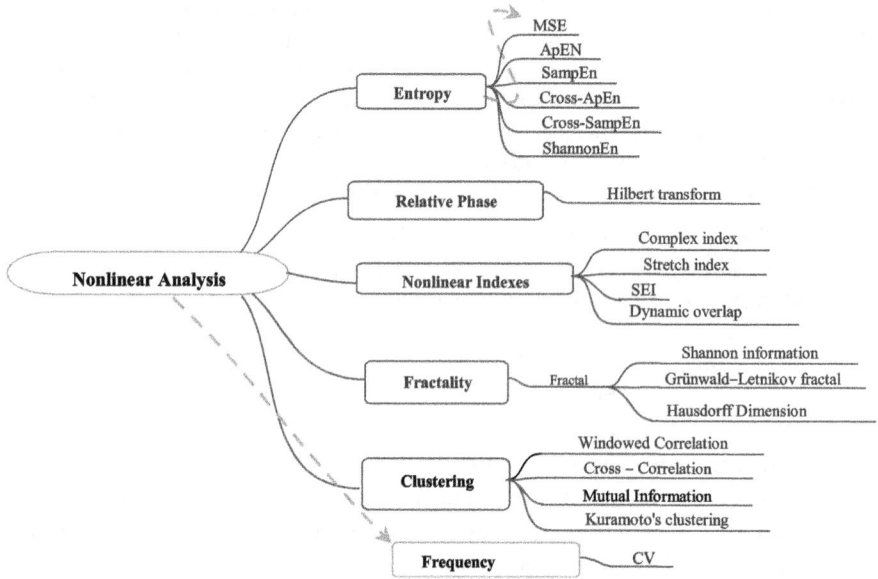

Figure 2.
Summary diagram for mathematical models and methodological procedures of the non-linear approaches.

$$CV(\%) = \frac{\sigma}{\bar{x}} \, 100 \qquad (23)$$

Speed synchronisation has also applied into a integrative approaches in some studies [10, 41]. The near-in-phase synchronisation to players' displacements is expressed in time spent (%) of time according to speed intensity zones: 0.0–3.5 km h^{-1} (low intensity); 3.6–14.3 km h^{-1} (moderate intensity); 14.4–19.7 km h^{-1} (high intensity); and >19.8 km h^{-1} (very high intensity) [41]. Summary diagram for mathematical models and methodological procedures of the non-linear approaches are presented in **Figure 2**.

Table 1 displayed the corresponding equation, thresholds, advantages, disadvantages and practical application for each nonlinear variable.

4. Practical considerations, criticisms and future perspectives

Matlab® routines (Math-Works, Inc., Massachusetts, USA) were the most selected procedure to analyse positional dataset in football. Universal Transverse Mercator (UTM) coordinate system were used to transform latitude and longitude data points [40, 66]. Methodological procedures differ on the correction guidelines to be used and reduce tracking signal noise, advising the use of 3 Hz Butterworth low pass filter [64, 92]. Several authors ran non-linear logarithms using 20 windows of 3000 points per data collect (i.e. ranged 5–25 Hz) [40, 67]. Integrating notational analysis and video-based tracking systems has been a worthwhile strategy to provide contextual significance to positional data [67, 68, 93]. Applying new analysis techniques based on big data still lack an integrative approach, and it will be interesting to understand how future studies can do so with techniques such as machine learning, deep learning or network analysis [13, 14]. These techniques have been extensively used to analyse positional and physiological variables, but there are still few studies under an integrative perspective [13, 36, 46]. There is a

Variable	Equation	Thresholds	Advantages	Disadvantages	Practical application
ApEn	$ApEn(m, r, N) = \phi^m(r) - \phi^{m+1}(r)$	$0 \leq ApEn \leq 2$; close 0—predictable; close 2—unpredictable	Similar patterns will not be followed by subsequent similar observations.	Dependent on the length record causing lacks of relative consistency.	Interpersonal coordination (1-vs-1 sub-phase); Opposition and cooperation relationships on collective movement behaviour.
SampEn	$SampEn(m, r, N) = -\ln \frac{\phi^{m+1}(r)}{\phi^m(r)}$	$0 \leq SampEn < \infty$; close 0—predictable; $< \infty$—unpredictable	Shorter time-series records with a greater relative consistency	Lower complexity for a signal than white noise signal.	Diagonal positioning of the goals on SSG
ShannonEn	$H(x) = \sum_{i=1}^{n} p_i \log_b p_i$ (Eq. 7)	$0 \leq H(x) \leq 1$; close 0—predictable; close 1—unpredictable	Multiple optimal weights on the evaluation and self-information.	Only consider a particular event, not the meaning of the events (criteria) themselves	Dynamics of tactical behaviour emerging on different time-scale using SSG
MSE	$Cross - SampEn(m, r, N) = \left\{ -\ln \frac{\phi^{m+1}(r)}{\phi^m(r)} \right\}$	$0 \leq Cross\text{-}SampEn < \infty$; close 0—predictable; $< \infty$—unpredictable	Faster and allows to evaluate two time-series crossed	Loss of pattern information hidden in the time series	Assessing the dynamics of team–team and player–team synchronisation
Hilbert transform	$\phi(t) = \phi_1(r) - \phi_2(r) = \arctan \frac{H_1 S_2(t) - H_2(t) S_1(t)}{S_1(t) S_2(t) + H_1(t) H_2(t)}$	In-phase ($-30° \leq \phi \leq 30°$) or anti-phase synchronisation ($-180° \leq \phi \leq -150°$ or $150° \leq \phi \leq 180°$); without synchronisation (other ϕ degree)	Require short signals than classical non-parametric methods	One dimensional processing causing phase ambiguities	Movement behaviour, speed synchronisation, inter-team distances, spatial interaction, oscillations of centroid position and surface area.
Dynamic overload	$<q_d(t)> = (1 - q_{stat})t^\alpha + q_{stat}$	$0 \leq q_d(t) \leq 1$; close 0—predictable; close 1—unpredictable	Compare dataset using a cosine auto-similarity that increase in each time lag	Analysis allowed the slow dynamics on a long timescale	Dynamical of tactical behaviour and constrained the perceptual-motor workspace
Stretch index	$s_{ind} = \frac{\sum_{i=1}^{N} w_i d_i}{\sum_{i=1}^{N} w_i}$	Near-in-phase synchronisation to players' displacements is expressed in time spent (%)	Provide the centroid position of the team and the sum of each player's dispersion on both axes	Relative stretch indexes has needed to measure two teams	Coordination and spatial interactions for opposite and team behaviours

Variable	Equation	Thresholds	Advantages	Disadvantages	Practical application
Windowed and cross correlation	$r_{xy} = \sum_{i=0}^{N-1} x_i y_i$	$-1 \leq r \leq 1$—phase (r close to -1), anti-phase synchronisation (r close to 1); without pattern (i.e., $r = 0$)	Measuring similarity, can determine time delay and the identity lagging signal	Bivariate linear association between group synchrony time-series data	Cross-correlation and peak picking for team synergies variability in tactical behaviour
Fractal Calcus	$I[P(x)] = -\log P(x)$ (Eq. (17))	0-Dimensional sets to 3-dimensional sets ($D = 0$–3)	Assessing fractal properties of human movement associated to sport skills and motor variability	Non-cyclicality of football movement using fractal analysis	Multifractal properties, dynamical stability and predictability of the movement.
Hausdorff dimension	$D(E) = \lim_{\varepsilon \to 0} \sup \frac{\log N_\varepsilon(\varepsilon)}{-\log \varepsilon}$				
Kuramoto's Clustering	$\rho_{group}(t_i) = \frac{1}{N} \sum_{i=1}^{n} \rho_{group,i}$	$0 \leq \rho_{group}$ or $\rho_{group,i} \leq 1$; close 0—predictable; close 2—unpredictable	Unbiased measure of group coordination and measure to assess player–team synchrony	Achieve synchronisation modes in networks with different structures	Order, disorder and variability in spatio-temporal interactions amongst two teams
AMI (clustering)	$I(x,y) = \sum_{X,Y} A P(x,y) \log\left(\frac{P(x,y)}{P(x)P(y)}\right)$	$0 \leq AMI < 1$; close 0—predictable; close 1—unpredictable	Measuring the nonlinear correlation of the two centroids' movements	Disadvantages of redundancy in each class	Positional synchronisation an geometrical center modifications in team behaviour
CV%	$CV(\%) = \frac{\sigma}{\bar{x}} 100$	NR	Statistical measure that is normalised and non-dimensional	Dependent on the mean values of the time-series	Speed synchronisation match-to-match variation

Abbreviations: AMI—*average mutual information*; ApEn—*approximate entropy*; CV—*coefficient of variation*; D—*dimension*; d_i—*distance between player i*; H—*Boltzmann–Gibbs–Shannon entropy*; $H_1S_2(t)$—*Hilbert transform*; I—*Shannon information*; m—*window length*; MSE—*multiscale entropy*; n—*frequency*; N—*time-series length*; NR—*not reported*; q_d—*dynamic overload*; q_{stat}—*horizontal asymptote*; r—*correlation*; r—*similarity radius or tolerance level*; SampEn—*sample entropy*; ShannonEn—*Shannon entropy*; SSG—*small-sided games*; τ—*time lag*; y_i—*data point*; τ—*timescales*; φ—*probability*; ρ_{group}—*group synchronisation*.

Table 1.
Summary of the non-linear variables and respective equation, thresholds, advantages, disadvantages and practical application.

lack of standardisation on non-linear measures, measurement and thresholds [20, 76]. It is even more evident in the physiological measures, therefore, the results obtained in studies that integrate positional and physiological datasets should be interpreted with caution [120]. The application of integrative approaches should also consider the boundaries between different key performance indicators such as the psychophysiological [45, 121–123], technical [44, 67, 93] and contextual factors [83, 84]. Also, acceleration outputs, metabolic power and body impacts have been poorly integrated with positional data. Behavioural data should still be better contextualised and the related-bias for physiological thresholds must be considered upon the time-dependent and transient reduction [84]. An integration approach to physiology and behavioural data must overcome some challenges on data visualisation, data processing (inherent to big data) and real-time tracking [13]. Moreover, futures researches should focus their analysis on women and sub-elite performers [20, 61].

5. Conclusion

Physiological assessment to monitoring training and match load has been carried out mainly under a linear perspective. Positional data to assess tactical behaviour considers fundamentally the theory of the complex systems and non-linear dynamics. Thus, an integrative approach allows a more holistic and extensive evaluation of the performance as a multifactorial phenomenon. This chapter summarises the theoretical concepts, mathematical models and methodological procedures to be applied by researchers and practitioners in training and match settings in football. The non-linear techniques reported more often in the literature were entropy, relative phase, complex indexes, correlation matrixes, clustering methods, frequency-based measures, fractals and multifractals. Correlation matrixes, clustering methods and fractality have not yet been applied in an integrative perspective in football. Finally, using non-linear approaches to integrate physiological and behavioural data remains a research-practice gap to be explored in the next years.

Acknowledgements

This research was supported by Portuguese Foundation for Science and Technology, I.P. (project UIDB04045/2021).

Conflict of interest

The authors declare no conflict of interest.

Author details

José E. Teixeira[1,2,3*], Pedro Forte[1,3,4], Ricardo Ferraz[1,5], Luís Branquinho[1,4,5], António J. Silva[1,2], Tiago M. Barbosa[1,3] and António M. Monteiro[1,3]

1 Research Centre in Sports, Health and Human Development, Covilhã, Portugal

2 Department of Sports, Exercise and Health Sciences, University of Trás-os-Montes e Alto Douro, Vila Real, Portugal

3 Department of Sport Science, Instituto Politécnico de Bragança, Bragança, Portugal

4 Department of Sports, Higher Institute of Educational Sciences of the Douro, Penafiel, Portugal

5 Department of Sports Sciences, University of Beira Interior, Covilhã, Portugal

*Address all correspondence to: jose.eduardo@ipb.pt

IntechOpen

References

[1] Silva P, Duarte R, Esteves P, Travassos B, Vilar L. Application of entropy measures to analysis of performance in team sports. International Journal of Performance Analysis in Sport. 2016;**16**(2):753-768

[2] O'Donoghue P. Research Methods for Sports Performance Analysis. London: Routledge; 2009

[3] Passos P, Araújo D, Volossovitch A. Performance Analysis in Team Sports. London: Routledge, Taylor & Francis; 2016

[4] Sarmento H, Clemente FM, Araújo D, Davids K, McRobert A, Figueiredo A. What performance analysts need to know about research trends in association football (2012–2016): A systematic review. Sports Medicine. 2018;**48**(4):799-836

[5] Barbosa TM, Goh WX, Morais JE, Costa MJ, Pendergast D. Comparison of classical kinematics, entropy, and fractal properties as measures of complexity of the motor system in swimming. Frontiers in Psychology. 2016;7:1566

[6] Duarte R, Araújo D, Folgado H, Esteves P, Marques P, Davids K. Capturing complex, non-linear team behaviours during competitive football performance. Journal of Systems Science and Complexity. 2013;**26**(1):62-72

[7] McGarry T, Anderson DI, Wallace SA, Hughes MD, Franks IM. Sport competition as a dynamical self-organizing system. Journal of Sports Sciences. 2002;**20**(10):771-781

[8] Kelso JAS, Schöner G. Self-organization of coordinative movement patterns. Human Movement Science. 1988;7(1):27-46

[9] Gréhaigne JF, Bouthier D, David B. Dynamic-system analysis of opponent relationships in collective actions in soccer. Journal of Sports Sciences. 1997;**15**(2):137-149

[10] Gonçalves B, Coutinho D, Travassos B, Folgado H, Caixinha P, Sampaio J. Speed synchronization, physical workload and match-to-match performance variation of elite football players. PLoS One. 2018;**13**(7):e0200019

[11] Duarte R, Araújo D, Davids K, Travassos B, Gazimba V, Sampaio J. Interpersonal coordination tendencies shape 1-vs-1 sub-phase performance outcomes in youth soccer. Journal of Sports Sciences. 2012;**30**(9):871-877

[12] Goes FR, Meerhoff LA, Bueno MJO, Rodrigues DM, Moura FA, Brink MS, et al. Unlocking the potential of big data to support tactical performance analysis in professional soccer: A systematic review. European Journal of Sport Science. 2021;**21**(4):481-496

[13] Rein R, Memmert D. Big data and tactical analysis in elite soccer: Future challenges and opportunities for sports science. Springerplus. 2016;5(1):1410

[14] Memmert D, Lemmink KAPM, Sampaio J. Current approaches to tactical performance analyses in soccer using position data. Sports Medicine. 2017;**47**(1):1-10

[15] Anderson BDO. Internal and external stability of linear time-varying systems. SIAM Journal on Control and Optimization. 1982;**20**(3):408-413

[16] Saberi A, Stoorvogel A, Sannuti P. Internal and External Stabilization of Linear Systems with Constraints. In Systems & Control: Foundations & Applications. Birkhäuser, Boston: Springer Science & Business Media; 2012

[17] Callier FM, Desoer CA. Linear System Theory. Springer Science & Business Media; 2012

[18] Vanrenterghem J, Nedergaard NJ, Robinson MA, Drust B. Training load monitoring in team sports: A novel framework separating physiological and biomechanical load-adaptation pathways. Sports Medicine. 2017;**47**(11): 2135-2142

[19] Impellizzeri FM, Rampinini E, Marcora SM. Physiological assessment of aerobic training in soccer. Journal of Sports Sciences. 2005;**23**(6):583-592

[20] Teixeira JE, Forte P, Ferraz R, Leal M, Ribeiro J, Silva AJ, et al. Monitoring accumulated training and match load in football: A systematic review. International Journal of Environmental Research and Public Health. 2021;**18**(8):3906

[21] Weiss L, Kalman RE. Contributions to linear system theory. International Journal of Engineering Science. 1965; **3**(2):141-171

[22] Peterson MD, Rhea MR, Alvar BA. Maximizing strength development in athletes: A meta-analysis to determine the dose-response relationship. Journal of Strength and Conditioning Research. 2004;**18**(2):377

[23] Busso T. Variable dose-response relationship between exercise training and performance. Medicine & Science in Sports & Exercise. 2003;**35**(7):1188-1195

[24] Goldstein EB. The ecology of J. J. Gibson's perception. Leonardo. 1981; **14**(3):191

[25] Davids K, Araujo D, Shuttleworth R, Button C. Acquiring skill in sport: A constraints-led perspective. International Journal of Computer Science in Sport. 2003;**2**:31-39

[26] Seifert L, Button C, Davids K. Key properties of expert movement systems in sport: An ecological dynamics perspective. Sports Medicine. 2013; **43**(3):167-178

[27] Stergiou N, Buzzi U, Kurz M, Heidel J. Nonlinear tools in human movement. In: Stergiou N, editor. Innovative Analyses of Human Movement. Champaign, IL: Human Kinetics; 2004

[28] Davids K, Button C, Araújo D, Renshaw I, Hristovski R. Movement models from sports provide representative task constraints for studying adaptive behavior in human movement systems. Adaptive Behavior. 2006;**14**(1):73-95

[29] Araújo D, Silva P, Davids K. Capturing group tactical behaviors in expert team players. In: Baker J, Farrow D, editors. Routledge Handbook of Sport Expertise. London: Routledge; 2015

[30] Travassos B, Davids K, Araujo D, Esteves P. Performance analysis in team sports: Advances from an ecological dynamics approach. International Journal of Performance Analysis in Sport. 2013;**13**:89-95

[31] Balague N, Torrents C, Hristovski R, Davids K, Araújo D. Overview of complex systems in sport. Journal of Systems Science and Complexity. 2013; **26**(1):4-13

[32] Kellert SH. Borrowed Knowledge: Chaos Theory and the Challenge of Learning across Disciplines. Chicago: University of Chicago Press; 2009. p. 303

[33] Ramos-Villagrasa PJ, Marques-Quinteiro P, Navarro J, Rico R. Teams as complex adaptive systems: Reviewing 17 years of research. Small Group Research. 2018;**49**(2):135-176

[34] Banerjee S. Chaos and complexity theory for management: Nonlinear dynamics. Hershey, PA: IGI Global; 2012

[35] Pikovsky A, Rosenblum M, Kurths J. Synchronization: A universal concept in nonlinear sciences. 2002;**70**:655-655

[36] Low B, Coutinho D, Gonçalves B, Rein R, Memmert D, Sampaio J. A systematic review of collective tactical behaviours in football using positional data. Sports Medicine. 2020;**50**(2): 343-385

[37] Clemente FM, Martins FML, Couceiro MS, Mendes RS, Figueiredo AJ. Developing a football tactical metric to estimate the sectorial lines: A case study. In: Murgante B, Misra S, Rocha AMAC, Torre C, Rocha JG, Falcão MI, et al, editors. Computational Science and its Applications – ICCSA 2014. Cham: Springer International Publishing; 2014

[38] Lames M, Erdmann J, Walter F. Oscillations in football - order and disorder in spatial interactions between the two teams. International Journal of Sport Psychology. 2010;**41**:85-86

[39] Pol R, Balagué N, Ric A, Torrents C, Kiely J, Hristovski R. Training or synergizing? Complex systems principles change the understanding of sport processes. Sports Medicine - Open. 2020;**6**(1):28

[40] Folgado H, Duarte R, Fernandes O, Sampaio J. Competing with lower level opponents decreases intra-team movement synchronization and time-motion demands during pre-season soccer matches. PLoS One. 2014;**9**(5):e97145

[41] Folgado H, Gonçalves B, Sampaio J. Positional synchronization affects physical and physiological responses to preseason in professional football (soccer). Research in Sports Medicine. 2018;**26**(1):51-63

[42] Sampaio JE, Lago C, Gonçalves B, Maçãs VM, Leite N. Effects of pacing, status and unbalance in time motion variables, heart rate and tactical behaviour when playing 5-a-side football small-sided games. Journal of Science and Medicine in Sport. 2014; **17**(2):229-233

[43] Ric A, Hristovski R, Gonçalves B, Torres L, Sampaio J, Torrents C. Timescales for exploratory tactical behaviour in football small-sided games. Journal of Sports Sciences. 2016;**34**(18): 1723-1730

[44] Canton A, Torrents C, Gonçalves B, Ric A, Salvioni F, Exel J, et al. The diagonal positioning of the goals modifies the external training load and the tactical behaviour of young football players. Biology of Sport. 2021;**39**(1): 135-144

[45] Ferraz R, Gonçalves B, Coutinho D, Oliveira R, Travassos B, Sampaio J, et al. Effects of knowing the task's duration on soccer players' positioning and pacing behaviour during small-sided games. International Journal of Environmental Research and Public Health. 2020;**17**(11):3843

[46] Carling C. Interpreting physical performance in professional soccer match-play: Should we be more pragmatic in our approach? Sports Medicine. 2013;**43**(8):655-663

[47] Bradley P, Ade J. Are current physical match performance metrics in elite soccer fit for purpose or is the adoption of an integrated approach needed? International Journal of Sports Physiology and Performance. 2018; **13**(5):656-664

[48] Baca A, Perl J. Modelling and Simulation in Sport and Exercise. London: Routledge; 2018

[49] O'Donoghue P. Sources of variability in time-motion data; measurement error and within player variability in work-rate. International Journal of Performance Analysis in Sport. 2004;**4**(2):42-49

[50] Anguera MT, Camerino O, Castañer M, Sánchez-Algarra P, Onwuegbuzie AJ. The specificity of observational studies in physical activity

and sports sciences: Moving forward in mixed methods research and proposals for achieving quantitative and qualitative symmetry. Frontiers in Psychology. 2017;8:2196

[51] Sánchez-Algarra P, Anguera MT. Qualitative/quantitative integration in the inductive observational study of interactive behaviour: Impact of recording and coding among predominating perspectives. Quality and Quantity. 2013;47(2):1237-1257

[52] Beato M, Coratella G, Stiff A, Iacono AD. The validity and between-unit variability of GNSS units (STATSports apex 10 and 18 Hz) for measuring distance and peak speed in team sports. Frontiers in Physiology. 2018;9:1288

[53] Rago V, Brito J, Figueiredo P, Costa J, Barreira D, Krustrup P, et al. Methods to collect and interpret external training load using microtechnology incorporating GPS in professional football: A systematic review. Res Sports Medicine. 2019;28 (3):437-458

[54] Gómez-Carmona CD, Bastida-Castillo A, Ibáñez SJ, Pino-Ortega J. Accelerometry as a method for external workload monitoring in invasion team sports. A systematic review. PLoS One. 2020;15(8):e0236643

[55] Leser R, Baca A, Ogris G. Local positioning systems in (game) sports. Sensors. 2011;11(10):9778-9797

[56] Ogris G, Leser R, Horsak B, Kornfeind P, Heller M, Baca A. Accuracy of the LPM tracking system considering dynamic position changes. Journal of Sports Sciences. 2012;30(14): 1503-1511

[57] Di Salvo C, Marco C. Validation of Prozone®: A new video-based performance analysis system.

International Journal of Performance Analysis in Sport. 2006;6(1):108-119

[58] Castellano J, Alvarez-Pastor D, Bradley PS. Evaluation of research using computerised tracking systems (Amisco and Prozone) to analyse physical performance in elite soccer: A systematic review. Sports Medicine. 2014;44(5):701-712

[59] Rico-González M, Los Arcos A, Nakamura FY, Gantois P, Pino-Ortega J. A comparison between UWB and GPS devices in the measurement of external load and collective tactical behaviour variables during a professional official match. International Journal of Performance Analysis in Sport. 2020; 20(6):994-1002

[60] Bastida-Castillo A, Gómez-Carmona CD, Sánchez EDLC, Pino-Ortega J. Comparing accuracy between global positioning systems and ultra-wideband-based position tracking systems used for tactical analyses in soccer. European Journal of Sport Science. 2019;19(9):1157-1165

[61] Teixeira JE, Forte P, Ferraz R, Leal M, Ribeiro J, Silva AJ, et al. Quantifying sub-elite youth football weekly training load and recovery variation. Applied Sciences. 2021;11(11): 4871

[62] Buchheit M, Allen A, Poon TK, Modonutti M, Gregson W, Salvo VD. Integrating different tracking systems in football: Multiple camera semi-automatic system, local position measurement and GPS technologies. Journal of Sports Sciences. 2014;32(20): 1844-1857

[63] Jackson BM, Polglaze T, Dawson B, King T, Peeling P. Comparing global positioning system and global navigation satellite system measures of team-sport movements. International Journal of Sports Physiology and Performance. 2018;13(8):1005-1010

[64] Sampaio J, Vitor M. Measuring tactical behaviour in football. International Journal of Sports Medicine. 2012;33:395-401

[65] Silva P, Chung D, Carvalho T, Cardoso T, Davids K, Araújo D, et al. Practice effects on intra-team synergies in football teams. Human Movement Science. 2016;46:39-51

[66] Folgado H, Lemmink KAPM, Frencken W, Sampaio J. Length, width and centroid distance as measures of teams tactical performance in youth football. European Journal of Sport Science. 2014;14(Suppl. 1):S487-S492

[67] Folgado H, Bravo J, Pereira P, Sampaio J. Towards the use of multidimensional performance indicators in football small-sided games: The effects of pitch orientation. Journal of Sports Sciences. 2019;37(9):1064-1071

[68] Olthof SBH, Frencken WGP, Lemmink KAPM. Match-derived relative pitch area changes the physical and team tactical performance of elite soccer players in small-sided soccer games. Journal of Sports Sciences. 2018; 36(14):1557-1563

[69] Ric A, Torrents C, Gonçalves B, Torres-Ronda L, Sampaio J, Hristovski R. Dynamics of tactical behaviour in association football when manipulating players' space of interaction. PLoS One. 2017;12(7): e0180773

[70] Carrilho D, Santos Couceiro M, Brito J, Figueiredo P, Lopes RJ, Araújo D. Using optical tracking system data to measure team synergic behavior: Synchronization of player-ball-goal angles in a football match. Sensors. 2020;20(17):4990

[71] Davids K, Hristovski R, Araújo D, Serre NB, Button C, Passos P. Complex Systems in Sport. London: Routledge; 2013

[72] Ric A, Torrents C, Gonçalves B, Sampaio J, Hristovski R. Soft-assembled multilevel dynamics of tactical behaviors in soccer. Frontiers in Psychology. 2016;7:1513

[73] Bannister EW, Clavert TW, Savage MV, Bach T. A systems model of training for athletic performance. Australian Journal of Sports Medicine. 1975;7:57-61

[74] Foster C, Florhaug JA, Franklin J, Gottschall L, Hrovatin LA, Parker S, et al. A new approach to monitoring exercise training. Journal of Strength and Conditioning Research. 2001;15(1): 109-115

[75] Coutts A, Kempton T, Crowcroft S, Coutts AJ, Crowcroft S, Kempton T. Developing athlete monitoring systems: Theoretical basis and practical applications. In: Kellmann M, Beckmann J, editors. Sport, Recovery and Performance: Interdisciplinary Insights. Abingdon, UK: Routledge; 2018

[76] Miguel M, Oliveira R, Loureiro N, García-Rubio J, Ibáñez SJ. Load measures in training/match monitoring in soccer: A systematic review. International Journal of Environmental Research and Public Health. 2021;18(5): 2721

[77] Carling C, Bloomfield J, Nelsen L, Reilly T. The role of motion analysis in elite soccer: Contemporary performance measurement techniques and work rate data. Sports Medicine. 2008;38(10): 839-862

[78] Bowen L, Gross AS, Gimpel M, Li F-X. Accumulated workloads and the acute: Chronic workload ratio relate to injury risk in elite youth football players. British Journal of Sports Medicine. 2017;51(5):452-459

[79] Gabbett TJ, Nassis GP, Oetter E, Pretorius J, Johnston N, Medina D, et al.

The athlete monitoring cycle: A practical guide to interpreting and applying training monitoring data. British Journal of Sports Medicine. 2017;**51**(20):1451-1452

[80] Nobari H, Fani M, Clemente FM, Carlos-Vivas J, Pérez-Gómez J, Ardigò LP. Intra- and inter-week variations of well-being across a season: A cohort study in elite youth male soccer players. Frontiers in Psychology. 2021;**12**:671072

[81] Aquino R, Carling C, Maia J, Vieira LHP, Wilson RS, Smith N, et al. Relationships between running demands in soccer match-play, anthropometric, and physical fitness characteristics: A systematic review. International Journal of Performance Analysis in Sport. 2020;**20**(3):534-555

[82] Vieira LHP, Carling C, Barbieri FA, Aquino R, Santiago PRP. Match running performance in young soccer players: A systematic review. Sports Medicine. 2019;**49**(2):289-318

[83] Paul DJ, Bradley PS, Nassis GP. Factors affecting match running performance of elite soccer players: Shedding some light on the complexity. International Journal of Sports Physiology and Performance. 2015;**10**(4):516-519

[84] Teixeira JE, Leal M, Ferraz R, Ribeiro J, Cachada JM, Barbosa TM, et al. Effects of match location, quality of opposition and match outcome on match running performance in a Portuguese professional football team. Entropy. 2021;**23**(8):973

[85] Gonçalves BV, Figueira BE, Maçãs V, Sampaio J. Effect of player position on movement behaviour, physical and physiological performances during an 11-a-side football game. Journal of Sports Sciences. 2014;**32**(2):191-199

[86] Bravi A, Longtin A, Seely AJ. Review and classification of variability analysis techniques with clinical applications. BioMedical Engineering OnLine. 2011;**10**:90

[87] Pincus SM. Approximate entropy as a measure of system complexity. Proceedings of the National Academy of Sciences of the United States of America. 1991;**88**(6):2297-2301

[88] Preatoni E, Ferrario M, Donà G, Hamill J, Rodano R. Motor variability in sports: A non-linear analysis of race walking. Journal of Sports Sciences. 2010;**28**(12):1327-1336

[89] Richman JS, Moorman JR. Physiological time-series analysis using approximate entropy and sample entropy. American Journal of Physiology-Heart and Circulatory Physiology. 2000;**278**(6):H2039-H2049

[90] Busa MA, van Emmerik REA. Multiscale entropy: A tool for understanding the complexity of postural control. Journal of Sport and Health Science. 2016;**5**(1):44-51

[91] Yentes JM, Hunt N, Schmid KK, Kaipust JP, McGrath D, Stergiou N. The appropriate use of approximate entropy and sample entropy with short data sets. Annals of Biomedical Engineering. 2013;**41**(2):349-365

[92] Duarte R, Araújo D, Correia V, Davids K, Marques P, Richardson MJ. Competing together: Assessing the dynamics of team–team and player–team synchrony in professional association football. Human Movement Science. 2013;**32**(4):555-566

[93] Coutinho D, Gonçalves B, Travassos B, Folgado H, Figueira B, Sampaio J. Different marks in the pitch constraint youth players' performances during football small-sided games. Research Quarterly for Exercise and Sport. 2020;**91**(1):15-23

[94] Grassberger P, Schreiber T, Schaffrath C. Nonlinear time sequence analysis. International Journal of Bifurcation and Chaos. 1991;**01**(03): 521-547

[95] Jamin A, Humeau-Heurtier A. (Multiscale) cross-entropy methods: A review. Entropy. 2020;**22**(1):45

[96] Balescu R. Equilibrium and nonequilibrium statistical mechanics. NASA STI/Recon Technical Report A. 1975;**76**:32809

[97] Shannon C. A mathematical theory of communication. Bell System Technical Journal. 1948;**27**(3):379-423

[98] Costa M, Goldberger AL, Peng C-K. Multiscale entropy analysis of biological signals. Physical Review E. 2005;**71**(2): 021906

[99] Palut Y, Zanone P-G. A dynamical analysis of tennis: Concepts and data. Journal of Sports Sciences. 2005;**23**(10): 1021-1032

[100] Gabor D. Theory of communication. Journal of the Institution of Electrical Engineers. 1943; **93**:429-457

[101] Frencken W, Lemmink K, Delleman N, Visscher C. Oscillations of centroid position and surface area of soccer teams in small-sided games. European Journal of Sport Science. 2011; **11**(4):215-223

[102] Hristovski R, Davids K, Araujo D, Passos P, Torrents C, Aceski A, et al. Creativity in sport and dance: Ecological dynamics on a hierarchically soft-assembled perception-action landscape. In: Davids K, Hristovski R, Araújo D, Balagué N, Button C and P. Passos, editors. Complex systems in sport. London: Routledge; 2013. pp. 259-271

[103] Clemente FM, Couceiro MS, Martins FML, Mendes R, Figueiredo AJ.

Measuring tactical behaviour using technological metrics: Case study of a football game. International Journal of Sports Science & Coaching. 2013;**8**(4): 723-739

[104] Travassos B, Gonçalves B, Marcelino R, Monteiro R, Sampaio J. How perceiving additional targets modifies teams' tactical behavior during football small-sided games. Human Movement Science. 2014;**38**: 241-250

[105] Derrick TR. Thomas JM. Time Series Analysis: The Cross-Correlation Function; 2004. p. 189

[106] Boker SM, Xu M, Rotondo JL, King K. Windowed cross-correlation and peak picking for the analysis of variability in the association between behavioral time series. Psychological Methods. 2002;**7**(3):338-355

[107] Silva P, Vilar L, Davids K, Araújo D, Garganta J. Sports teams as complex adaptive systems: Manipulating player numbers shapes behaviours during football small-sided games. Springerplus. 2016;**5**(1):191

[108] Moura FA, van Emmerik REA, Santana JE, Martins LEB, Barros RML de, Cunha SA. Coordination analysis of players' distribution in football using cross-correlation and vector coding techniques. Journal of Sports Sciences. 2016;**34**(24):2224–2232

[109] Sparrow WA, Donovan E, van Emmerik R, Barry EB. Using relative motion plots to measure changes in intra-limb and inter-limb coordination. Journal of Motor Behavior. 1987;**19**(1):115-129

[110] Freitas Cruz I, Sampaio J. Multifractal analysis of movement behavior in association football. Symmetry. 2020;**12**(8):1287

[111] Couceiro MS, Clemente FM, Martins FML, Machado JAT. Dynamical

stability and predictability of football players: The study of one match. Entropy. 2014;**16**(2):645-674

[112] Lopes AM, Tenreiro Machado JA. Entropy analysis of soccer dynamics. Entropy. 2019;**21**(2):187

[113] Couceiro MS, Clemente FM, Martins FML. Analysis of football player's motion in view of fractional calculus. Central European Journal of Physics. 2013;**11**(6):714-723

[114] Ramirez-Arellano A, Bermúdez-Gómez S, Hernández-Simón LM, Bory-Reyes J. D-summable fractal dimensions of complex networks. Chaos, Solitons & Fractals. 2019;**119**:210-214

[115] Rokach L, Maimon O. Clustering methods. In: Maimon O, Rokach L, editors. Data Mining and Knowledge Discovery Handbook. Boston, MA: Springer US; 2005

[116] Frank TD, Richardson MJ. On a test statistic for the Kuramoto order parameter of synchronization: An illustration for group synchronization during rocking chairs. Physica D: Nonlinear Phenomena. 2010;**239**(23): 2084-2092

[117] Kuramoto Y, Nishikawa I. Statistical macrodynamics of large dynamical systems. Case of a phase transition in oscillator communities. Journal of Statistical Physics. 1987; **49**(3):569-605

[118] Cover TM, Thomas JA. Elements of Information Theory. New York: Wiley; 1991. p. 542

[119] Silva P, Duarte R, Sampaio J, Aguiar P, Davids K, Araújo D, et al. Field dimension and skill level constrain team tactical behaviours in small-sided and conditioned games in football. Journal of Sports Sciences. 2014;**32**(20): 1888-1896

[120] Gonçalves B, Coutinho D, Exel J, Travassos B, Peñas C, Sampaio J. Extracting spatial-temporal features that describe a team match demands when considering the effects of the quality of opposition in elite football. PLoS One. 2019;**14**:e0221368

[121] Ferraz R, Gonçalves B, Coutinho D, Marinho DA, Sampaio J, Marques MC. Pacing behaviour of players in team sports: Influence of match status manipulation and task duration knowledge. PLoS One. 2018;**13**(2): e0192399

[122] Branquinho L, Ferraz R, Travassos B, Marques MC. Comparison between continuous and fractionated game format on internal and external load in small-sided games in soccer. International Journal of Environmental Research and Public Health. 2020;**17**(2): 405

[123] Branquinho L, Ferraz R, Travassos B, Marinho DA, Marques MC. Effects of different recovery times on internal and external load during small-sided games in soccer. Sports Health. 2021;**13**(4):324-331

Justification of some Aspects of the Choice of Training Means Selection in Track-and-Field Jumps

Mikhail Shestakov and Anna Zubkova

Abstract

The chapter deals with the aspects of a take-off in track-and-field jumps with regard to biomechanics and physiological processes. In this chapter, we describe biomechanical and physiological processes underlying the main biomechanisms (BM), which are involved in track-and-field jumps. Our investigation aims at confirmation of the hypothesis that the concept of BM forms the basis of the approach to selecting technique development means in track-and-field. The aim of the first part of the research was to compare the contribution of different BMs. We have analyzed biomechanical parameters of the take-off in a group of elite jumpers (n = 50) during official competitions. Computer simulation modeling was used to detect how an increase in the run-up speed changed the contribution of different BMs. The aim of the second part of the research was to examine the peculiarities of a take-off in special exercises. Findings of the research demonstrated that the take-off in training exercises was performed using relatively independent BMs, similar to those used in competitive jumps. Being dependent on the motor task, key biomechanisms appear to be interdependent on the dynamic level. The role and contribution of the BMs depend on the type of exercise or conditions of its execution, initial conditions, and a motor task set to an athlete.

Keywords: track-and-field jumps, muscle-tendon unit, biomechanism, special exercises

1. Introduction

At present, having taken a systemic-structural approach, general biomechanics studies feature of the locomotor system of a man, biomechanical characteristics of movements, composition, and structure of motions in sports exercises and movements.

This approach suggests to single out spatial and temporal elements in a system of movements [1]. Just here, we see a contradiction, as a material system cannot be subdivided into spatial and temporal elements. From the point of view of contemporary philosophy of scientific cognition, it is incorrect to think of processes, properties, or relations as being systems. They all are no more than manifestations of various properties of a material object, while a system is a model of an original material object, the latter also consisting of material elements.

An approach used by J.G. Hay [2] is the most recognized in sport biomechanics now. Its essence (illustrated by a vertical standing jump) consists in subdivision of the trajectory of the body center of gravity (COG) into a few segments. The following identification of biomechanical characteristics responsible for the COG displacement and velocity of displacement is based on common sense. For example, arm lift-up shifts the COG upwards; legs extension produces the same effect; legs length ensures a certain position of the body COG at the end of take-off. On the ground of common sense, important parameters are selected and subjected to correlation analysis in order to find their relationship and to obtain multiple regression equations. The logic of this approach is similar to that of an empirical research aimed at formulation of an empirical law, which does not reveal the essence of a phenomenon, although enables to make some suggestions.

In biomechanics, we may be unfamiliar with brain organization and the central nervous system (CNS) can be considered as a "black box." Here lies the boundary between physiology and biomechanics. A biomechanist must be proficient in programming deliberate motor actions aimed at reaching a preset goal, i.e. motor programs. Physiological concepts of movement control do not substantiate the laws of mastering motor actions. However, to achieve success, a coach needs knowledge both in biomechanics and physiology in order to create training programs aimed at the development of motor programs of a competitive movement in an athlete. Thus, to work out a plan of technique development of a track-and-field athlete, it appears necessary to model a competitive movement, as well as training means to be used besides the principle movement.

To model the locomotor system of a man, we must use ideal models from theoretical mechanics [3]. Theoretical mechanics use models including the following elements: two- or three-dimensional space, time, point mass, perfectly rigid body (a rod), hinge, kinematic chain, ideal liquid or gas, etc. [4]. All these models are used in biomechanics, although to create an adequate model of a locomotor system, a model of muscle is needed. Hence, the subject of biomechanics matches that of theoretical mechanics only partially.

At every single instant, human existence can be considered as a combination of biomechanisms. In general, the concept of biomechanism includes biochemical objects (mitochondria, myofibrils, etc.), physiological systems (cardiovascular, endocrine, immune, central nervous, and other systems), and, the locomotor system.

In biomechanics, this concept should be referred to mechanics, in particular, to the theory of machines and mechanisms. Let us define a biomechanism (BM) as an aggregate of certain body parts movements, independent of other parts movements, transforming one type of energy into another that leads to changes in the position and speed of the athlete's body COG while accomplishing a certain movement task in certain external conditions [5–7].

To control a multilink system, the CNS combines separate links into subsystems (key kinematic mechanisms), which can act independently, although in doing so to pursue a common goal. A biomechanism as an integral system consists of a set of components, each possessing its own properties, which can manifest themselves in human movements in different ways.

The following components are singled out in a biomechanism:

1. Muscle as:

   • a converter of chemical energy into mechanical one;

   • a resilient element, capable of storing and returning energy;

• a ductile element, capable of damping external loads;

• an energy (power) transmitter from energy sources.

2. Bone as:

• a lever for force and power transfer;

• a pendulum for energy conversion;

• a rod for support and reaction against external loads.

3. Joint as:

• a hinge joining bones in a kinematic chain;

• a hinge restricting mobility of joined bones.

Besides that, we should take into account controlling units containing control programs (motor programs that are formed, stored, and functioning in the athlete's CNS) [8, 9].

It is noteworthy that a CNS model must meet strict requirements. The model must reflect the process of controlling the object (in our case, the locomotor system) as well as model environment conditions and their relationship. Important is for both processes to be modeled as parallel [10]. The ability to perform deliberate movements means that a person can control target-oriented movements of the body or its parts more or less precisely. As the purpose of a movement is supposed to be solved by a certain BM and perceived by consciousness, it can be controlled and changed deliberately. The problem of movement spatial & temporal parameters differentiation, i.e. a method of performing a motor action, or, in other words, technique of a movement is, probably, solved by means of conscious control of certain BMs [6].

Hypothesis. Our investigation aims at confirmation of the hypothesis that the concept of bimechanism (BM) forms the basis of the approach to selecting technique development means in track-and-field. We examined one of the most important components of track-and-field jumps, i.e. a take-off.

The following BMs can be identified in the take-off in track-and-field jumps:

• BM of the support leg and body extension (LBE);

• BM of the arms and swinging leg swinging motion (SM);

• BM of the "overturned pendulum" (OP);

In the publication [11] the authors underscore three factors that play a key role in the biomechanism of the support leg and body extension. They are:

• consecutive extension of the coxofemoral and knee joints;

• differently directed changing angles in the coxofemoral and knee joints in the transition phase from shock absorption to take-off;

• optimal legs bending in the knee joints.

Swinging motions contribution. This mechanism increases the vertical component of the COG velocity after take-off [12]. It ensures:

- growth of the support reaction force due to the accelerated motion of the swinging links [13];

- growth of the swinging links velocity till the start of the knee joint extension [14];

- correct position of the swinging links at the end of take-off [13].

The essence of the "overturned pendulum" biomechanism consists in the ability to increase the COG vertical velocity due to the athlete's body pivoting over the point of bearing [15, 16].

Relatively independent kinematic mechanisms are interdependent at dynamic level, i.e. realization of any of them affects the efficiency of the others. The role and contribution of the key kinematic mechanisms to the result demonstrated by an athlete depend on the type of a jump, initial conditions, and the task set. There exist different ways of realization of any BM as well as different interaction between BMs within the same jumping event.

Physiological basis of biomechanisms (BMs).

To study a concept of "biomechanism" (BM) in voluntary movements, we should understand physiological processes which take place in muscles directly involved in the biomechanisms.

An efficient take-off in jumps depends on well-coordinated simultaneous activation of three biomechanisms:

- BM of the "overturned pendulum" that helps additional to load muscles-extensors of the knee joint of the support leg;

- BM of "swinging motion of the arms and swinging leg" that provides additional load on the support leg joints and muscles;

- BM of the "support leg and body extension" that provides elastic energy storage in ankle muscles-flexors (plantar extensors) of the support leg.

The BM of the "support leg and body extension" involves more muscle activity than the other two BMs ("overturned pendulum" and "swinging motion of the arms and swinging leg"), which play complementary roles by increasing the impact of the first one on the muscles.

In track-and-field jumps, athletes try to develop the maximal power of movement. Greater power of muscle contraction can result from an increase in either its strength, or its velocity, or both components. As a rule, the most significant gain in power is due to the increase of muscular strength.

Multiple studies demonstrate that manifestation of greater power in jumps may be related to physiological peculiarities of muscle-tendon unit (MTU) activity [17]. Elastic energy of tendons is used for solving various motor tasks, notably for increasing output power of the muscle-tendon unit (MTU) [7].

There are two ways to increase the efficiency of power development in the BM of the "support leg and body extension": pre-stretch of skeletal muscles and mechanical energy transfer via biarticular muscles.

As regards the first way of increasing power of a take-off (pre-stretch of skeletal muscles), it is known that pre-stretch of muscle-tendon unit (MTU) enhances strength of its subsequent contraction [18]. This mechanism can only be involved if extension starts from the most powerful hip joint due to m.gluteus maximus (GM) contraction. The energy is transferred from the hip joint to the shin via m. rectus femoris (RF) m. gastrocnemius (GA) is not involved into work immediately after the extension of the knee joint; it requires some time for GA to start contraction. Elastic energy is stored in the tendons of ankle extensors and then released quickly. As a result, the ankle joint being controlled by the weakest muscles can develop greater power due to the energy transferred from the proximal joint [19]. Thus, the maximal power of muscle contraction is achieved due to conjoint activity of muscles and tendons involved in the mechanism of energy transfer based on the pre-stretch effect.

The second way of increasing power of a take-off (energy transfer via biarticular muscles) was described in a few works including those related to jumps [17, 19]. It also involves processes which take place in the MTU. This mechanism is based on the fact that powerful monoarticular gluteus muscles, in particular GM contribute most to hip extension, thus increasing the angle in the hip joint. The energy is transferred from GM to the knee joint via the biarticular RF, and the knee joint is extended by RF and a group of monoarticular m. vastus (VA). Energy transfer via a biarticular muscle takes place when the muscle contracts, although it develops the greatest power when working in almost isometric regimen, i.e. contracting very slowly [20]. At the same time, knee extension causes plantar flexion in the ankle joint due to energy transfer via m. gastrocnemius (GA) that reinforces contraction of the triceps muscle of the calf [21].

The description of two ways to enhance power of a take-off shows that the MTU pre-stretch requires less time than energy transfer via biarticular muscles. In some works, the first way of power enhancement is called a "catapult" [22, 23]. This time difference is very important with regard to interaction of the BM of "support leg and body extension" with the other two BMs. In the first case (MTU pre-stretch) greater contribution of the "overturned pendulum" will decrease the effectiveness of the mechanism, whereas greater contribution of the "swing" will have positive effect on the result. Greater velocity of the swing owing to active contraction of the working muscles will lead to greater storage of elastic energy, greater stiffness of the support leg, and consequently greater power of the movement. Meanwhile shorter time of the movement will decrease the effect of MTU pre-stretch. In the second case (energy transfer via biarticular muscles) extra loading of the support leg due to greater contribution of the "overturned pendulum" will have positive effect. Greater contribution of the "overturned pendulum" is reached by increasing the step length and thereby lowering the COG trajectory that increases time of the take-off.

As a practical matter, this knowledge is important for correct planning and training. Training exercises should be performed by athletes taking into account their individual peculiarities based on different ratios of BMs contribution in the take-off power.

2. Objective and methods

We have analyzed biomechanical parameters of the take-off in a group of elite Russian male jumpers (n = 50) in competitive conditions (during official contests). The aim of this part of the research was to compare the contribution of different BMs involved into take-off in track-and-field jumps. Video recording was made by a digital camera JVC-9800 with the speed of 50 frames per second. Having been captured by standard computer programs, the image of the jumper's body was modeled by virtue of an anthropomorphic 12-segment [5]. The computer complex consisted of a few modules: calculation of mass-inertial parameters of an athlete;

calculation of kinematical and energy characteristics of movements of separate body links and the whole body based on videotape processing (it allowed to determine linear and angular indices of the body links kinematics as well as potential, kinetic and full energy of each link). The unique feature of this module is the capability to determine changes in length and contraction speed of 9 major muscles of the lower extremities. It permits to determine key peculiarities of the athlete's technique and to simulate conditions, under which top results could be achieved. Mathematical processing was done in the Scientific Research Institute of the Russian State University of Physical Education, Sport, and Tourism. The accuracy of measurements was determined in a metrological study and accounted for 0.01 m (linear parameters) and 0.02 mps (velocity parameters).

3. Results of the first part of the research

Having analyzed the results, we determined the contribution of BMs into take-off in jumps regarding changes in the full energy of separate segments and links of the athlete's body. Simulation modeling enabled us to make conclusions concerning not only the ratio of BMs contribution into take-off in every jump event and a comparison of different jumps (horizontal and vertical) but to monitor the change of BMs contribution into take-off in case of 5% increase of the athlete's COG velocity at the last step of the approach run in each type of jumps and its effect on the result.

Results demonstrated by the athletes in the course of the experiment attained the level of master of sport, international class (**Table 1**). The simulation modeling showed that the growth of the approach speed would provide real chances of getting into the World championship finals.

Table 2 displays proportion (in %) of the BMs contribution into take-off based on changes in the full energy of the athlete's body links measured in the experiment.

The greatest contribution of the BM of swinging links into the take-off is at once apparent. The contribution of the BM of the take-off leg extension is more pronounced in pole vault and less pronounced in high jump, whereas the contribution of the BM of "overturned pendulum" is greater in high and long jumps.

High and long jumps are the most similar in what concerns the structure of BMs operation at take-off, although the execution of the jumps is quite different (horizontal and vertical directions).

Parameter	Long jump	High jump	Triple jump	Pole vault
real result (at the registered COG velocity), cm	805 ± 12	228 ± 2.3	1715 ± 35	565 ± 5.5
Simulated result (at the modeled COG velocity), cm	820	232	1740	575

Table 1.
Jumping results: Real and obtained by simulation modeling.

Biomechanism	Long jump (%)	High jump (%)	Triple jump (%)	Pole vault (%)
Leg and body extension	15.8	14.3	16.1	18.9
Swinging links	69.8	70.2	70.5	67.9
Overturned pendulum	14.4	15.5	13.4	13.2

Table 2.
Contribution of different BMs into take-off in track-and-field jumps (%).

We decided to find out what would happen if the athlete's speed at the last step of the approach run (just prior to take-off) increased by 5%. We took the value 5% because the examination of strength-velocity qualities of top-class athletes permitted to suppose such an increase of speed to be attainable by advanced jumpers, who seem to be capable of bearing higher strength loads.

An increase in speed will naturally lead to an increase in the body links energy. The question arises if the energy growth will be proportional to the speed of the COG in every BM.

Important changes were observed in the BM of swinging links (**Figure 1**). In high and long jumps the contribution of this BM not only increased, but started earlier. As for triple jump, both temporal and amplitude parameters of the BM of swinging links grew. In pole vault, the increase was proportional to the increase of the COG velocity. The structure of the BM of the take-off leg extension for all the jumps under study remained the same. Considerable changes took place in the BM of "overturned pendulum" in high and long jumps. Its contribution became more important in triple jump but nearly did not change in pole vault.

Changes in the ratio (in %) of BMs contribution into take-off are shown in **Table 3**.

We should also note an increase in the total energy of the BMs, the greatest being observed in high and triple jumps.

On present evidence, it may be suggested that high and long jumps are the most similar in the structure of take-off, lesser similarity being found with triple jump and pole vault. Therefore, horizontal and vertical jumps reveal a certain resemblance in the structure of take-off (**Table 4**).

The greatest increase in the body COG speed can be reached by intensifying movements of swinging links (as in amplitude, as in temporal parameters). At the same time, training means and methods aimed at the development of the BM of the take-off leg and body extension should not be excluded from training programs, because the links of this BM would have to bear the increased loads resulting from

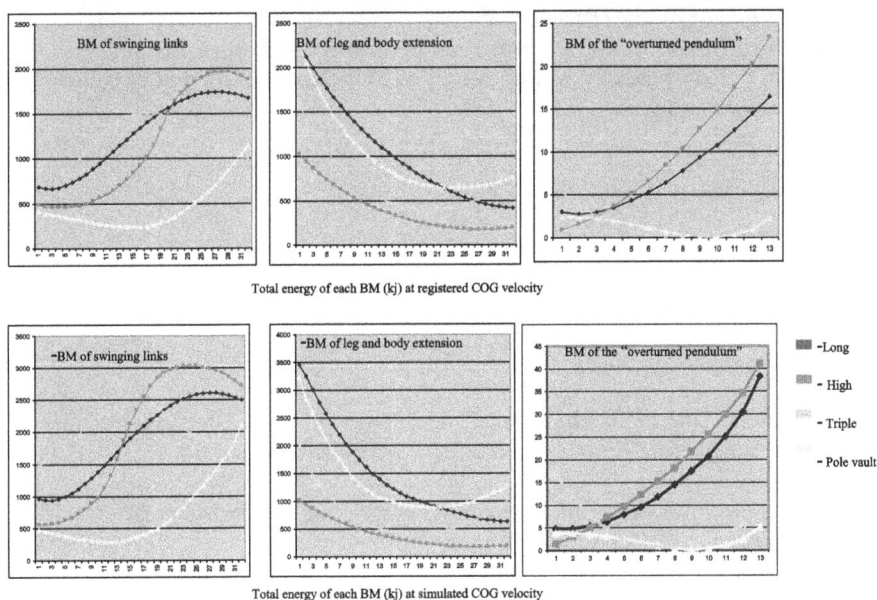

Figure 1.
Change of BMs contribution to take-off.

Biomechanism	Long jump (%)	High jump (%)	Triple jump (%)	Pole vault (%)
Leg and body extension	13.5	13.7	12	18.7
Swinging links	71.5	72.6	75.5	68.9
Overturned pendulum	15.0	14.3	12.5	12.4

Table 3.
Changes in BMs contribution into take-off obtained by modeling a 5% increase in velocity at the last step of the approach.

Biomechanism	Long jump (%)	High jump (%)	Triple jump (%)	Pole vault (%)
Leg and body extension	53	52.2	52	51.9
Swinging links	56.6	61.81	56.59	63.9
Overturned pendulum	57.9	59.25	57.67	54.1

Table 4.
Changes of the total energy in every BM (in %) obtained by modeling a 5% increase of speed at the last step of the approach.

more intensive work of the swinging links. This will provide efficient work of the BM of "overturned pendulum".

Data obtained in this part of the research permit to conclude that the structure of take-off in track-and-field jumps is formed according to a certain motor objective that depends on the character of the jump.

The speed of the athlete's body COG at take-off in all jumping events is controlled to a great extent by swinging movements of the body links, in other words, it depends on amplitude and temporal parameters of the swing.

3.1 Pedagogical requirements to training means selection

In running jumps, the efficiency of key kinematic mechanisms and, consequently, the efficiency of the athlete's interaction with the support depends on movements pattern executed by an athlete and aimed at realization of the following pedagogical requirements:
For all jumps:

• to run as fast as possible in the approach;

• to start active swinging movements before touch-down.

For taking off in long and high jumps:

• to lower the body COG at the last 2–3 approach strides;

• to plant the take-off foot by a "paddling" movement, but at less angle with the support (i.e. in front of the body) and to lean the trunk backward from the vertical.

The degree of realization of some of the requirements differently affects the realization of the others. For example, the COG lowering at the last approach strides in high and long jumps (contrary to triple jump and pole vault) creates favorable conditions for correct placement of the take-off foot and less angle of the body lean with the support, in spite of setting higher requirements for strength-velocity qualities of the take-off leg muscles. This should increase the contribution of the "overturned pendulum" and swinging links BMs in the phase of the body and take-off leg pivoting upon the point of support so that the trunk becomes positioned vertically above it. The subsequent forward-downward rotation of the take-off leg causes lowering of the knee and coxofemoral joints that is compensated by the mechanism of the take-off leg and body extension. When the take-off leg is planted "under" the trunk at take-off (a popular method of learning technique in long jumps), the take-off leg is lowered (forward-downward rotation), that can be considered as a technique fault, because in this case the efficiency of the BM of the take-off leg and body extension and the contribution of the "overturned pendulum" decrease.

The realization of some of the requirements mentioned above does not always favor the realization of the others. For instance, a too fast approach increases high-impact and inertia loads on the take-off leg, in particular, when it is placed at narrow angle with the support.

Taking into account the take-off structure in a given type of jumps and the peda-gogical requirements listed above, any training program in track-and-field should include special means, each affecting technical skills depending on the core and form of a certain movement.

The revealed phenomena allowed to set objectives for the second part of the research aimed at biomechanical investigation of training means most frequently used for technique development in track-and-field jumping events.

To examine biomechanical features of take-off in special exercises primarily used in technique development sessions by track-and-field jumpers (in different jumping events), we have carried out a laboratory experiment on a special complex "Qualisys" (Sweden) using high-speed recording camera (240 frames per second). 5 elite male track-and-field jumpers, regularly performing in international competi-tions took part in the experiments (2 long jumpers, 2 high jumpers, 1 pole-vaulter). The age of the subjects was 22 ± 1.4 yrs., duration of practicing track-and-field jumps 7 ± 2.3 yrs., height 1.84 ± 0.04 m, weight 74 ± 3.1 kg. Exercises performed in the experiment included: long jump, long jump over a hurdle (0.96 m) placed in 1 m distance from the take-off spot, long jump taking off a raised or lowered board (0.05 m), jump up with touching an object suspended at 2.5 m height and in 1 m distance from the take-off spot, a pattern of 3 hops after an approach run. All exercises were performed after 6 running strides at the maximal approach speed.

4. Results of the second part of the research and their discussion

4.1 Biomechanism of the take-off leg and body extension

In jumps, the greatest mechanical impact directed at stretching biarticular muscles of the lower extremities, in particular, rectus femoris, is achieved at take-off due to simultaneous forced bending of the take-off leg in the knee joint (at shock absorption) and its active straightening in the coxofemoral joint. In this case, the tractive force produced by this muscle is aimed at the knee joint extension. This element of the BM of the take-off leg and body extension at the phase of interaction with the support outwardly looks as the leg flexion with the simultaneous driving of the pelvis and knees forward while leaning backward.

Biceps femoris has two functions in running jumps. According to our data and that reported in other studies, the take-off leg hits the take-off board at the angle of 59–74° with the horizontal, the angles in the coxofemoral and knee joints vary-ing within the range of 165–170° and 160–175° correspondingly, depending on a jump event (**Table 5**). The trunk having "run against" the support leg planted on a

Exercise	Angle in the ankle joint (AJ) at touch-down	Angle in the knee joint (KJ) at touch-down	Angle in the coxofemoral joint (CFJ) at touchdown
Long jump, °	67	169	171
Long jump over a hurdle	65	172	175
Long jump from a raised take-off platform (0.05 m), °	60	165	170
Long jump from a lowered take-off platform (0.05 m), °	74	174	168
Jump up with touching an object by hand, °	59	175	172
pattern of 3 hops, °	70	160	165

Table 5.
Kinematic characteristics of take-off in exercises under study (°).

take-off board, starts pivoting forward. Less angles of the support leg touch-down and body lean (measured clockwise with respect to the horizontal) were observed in the following exercises: jump up with touching a suspended object by hand, long jump from a raised platform (0.05 m).

The analysis of the dynamics of αCFJ/αKJ (**Figure 2**) and values of angles in the involved joints in different types of jumps showed that the speed of contraction in biarticular muscles is lower than in those being monarticular, and consequently, the tractive force produced by biarticular muscles is greater.

It led us to suggest that biarticular muscles play more significant role in providing efficient interaction with the support in jumps. In this context, training means and exercises should be selected so that they could develop strength-velocity qualities of those muscles in plyometric regimen of contraction, primarily with oppositely directed change in angles (as in the pairs CFJ-KJ and KJ-AJ).

Maximal results in jumps are reached when the angle of the knee joint flexion at shock absorption is optimal. These optimal values are different for different jump events. Similar in all the jumps is that the amplitude of the forced leg bending varies within the range of 25–35° and is independent of the type of a jump.

The examination of the three key features of the BM of legs and body extension demonstrated that at dynamic level the structure of the locomotor system determined the specificity of interaction with support in jumping exercises. Under otherwise equal conditions, the following factors are thought to be the most important:

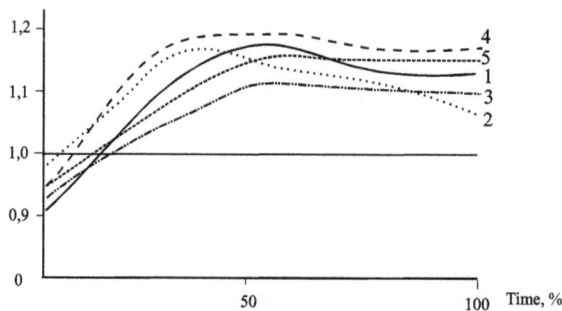

Figure 2.
Changes in angles in the coxofemoral and knee joints (αCFJ/αKJ) of the take-off leg at take-off. 1- long jump, 2 - long jump over a hurdle; 3- long jump from a raised take-off platform (0.05 m), 4 - long jump from a lowered take-off platform (0.05 m), 5 -jump up with touching an object by hand °.

1) maximal values and ratio of force momentums in the joints being involved, and
2) plyometric regimen of contraction of monarticular and, in particular, biarticular
muscles.

4.2 Swinging links motion

Additional stretching of the lower extremities muscles at the end of shock absorption is provided by external mechanical load originating from the vertical component
of inertia forces (Fin) applied to the centers of mass of swinging links and transferred
to the centrifugal force (Fcf), directed along the kinematic chain. The value of Fin
depends as on the swinging movement of swinging links, as on the accelerated lift of
the linkage points of those links: for arms – the shoulder girdle lift and trunk straightening; for the swinging leg – lift of the pelvis (due to the take-off leg straightening and/
or trunk pivoting over the point of bearing in accordance with the BM of the "overturned pendulum"). Contribution of the accelerated lift of the linkage points of the
swinging links can be estimated from the difference between the values of Fin and Fcf.
The links are accelerated by:

• positive force momentums in the shoulder joints and the coxofemoral joint of
the swinging leg;

• a decrease of the swinging links radius of inertia (arms flexion in the elbow
joints and the swinging leg flexion in the knee joint). According to the law of
kinetic momentum conservation, it leads to an increase in the angular velocity
of rotating links.

Deceleration of the swinging links goes on in the reverse order – the radius of
inertia grows and the sign of the force momentum changes from positive to negative
one due to the action of antagonist muscles. This enables an abrupt reduction of Fin
in the centers of mass of the swinging links up to zero that, consequently, reduces
the load on the lower extremities muscles at the end of the transfer from the plyometric regimen of their contraction to the myometric one. It is the effect of a sudden release of a stretched active muscle [24] Therefore, at that instant the swinging
links should have gained the maximal momentum in the direction of the take-off,
and the lower extremities should work on the acceleration of the trunk solely.

Results displayed in **Table 6** show that inertia forces in swinging motions caused
significant changes in the COG vertical velocity at take-off, which were due to:
creation of additional load on muscles-extensors of the lower extremities at the
end of shock absorption phase (inertia forces being transferred to the support by
kinematic chains);

Exercise	Fin, swinging leg	Fin, upper extremities
Long jump	1253 ± 54	655 ± 24
Long jump over a hurdle	1080 ± 36	583 ± 35
Long jump from a raised take-off platform (0.05 m)	1345 ± 44	674 ± 64
Long jump from a lowered take-off platform (0.05 m)	1437 ± 87	812 ± 75
Jump up touching a suspended object by hand	1315 ± 89	702 ± 72
pattern of 3 hops	1214 ± 65	733 ± 32

Table 6.
Maximal vertical component of inertia forces of the swinging links centers of mass at take-off (N).

Exercise	Shift of the marker, mm
Long jump	1.2 ± 0.09
Long jump over a hurdle	4.6 ± 0.55
Long jump from a raised take-off platform (0.05 m)	6.1 ± 1.4
Long jump from a lowered take-off platform (0.05 m)	3.0 ± 0.8
Jump up touching a suspended object by hand	5.1 ± 1.3
pattern of 3 hops	−2.1 ± 0.07

Table 7.
Shift of the marker placed at the CFJ axis of rotation of the take-off leg at shock absorption.

growth of the swinging links velocity until the start of the knee joint extension; swinging links position at the end of take-off.

4.3 Biomechanism of the "overturned pendulum"

According to our data, the highest (0.06 m) lift of the pelvis (or the marker attached at the point of the CFJ axis of rotation of the take-off leg) at shock absorption takes place in long running jumps with the take-off from a raised platform, despite the knee joint flexion.

It was found out that in a hop performed after an approach run the center of mass of the take-off leg thigh was raised by 0.03 m, while the motion of the swinging links was directed forward-downwards. As consequence of these compensatory movements in long jumps the body COG moves in parallel to the support, and in triple jump (step and jump phases) the body COG is lowered toward the support (**Table 7**).

The evidence concerning the BM of the "overturned pendulum" proved that its efficiency depends to a certain extent on the position of the athlete's body at touch-down. The less is the touch-down leg angle and the more is the body backward lean, the longer will the distance used for accelerating the pelvis, trunk, and the whole body be.

5. Practical recommendations

Findings of the second part of the research demonstrate that there exist specific biomechanical characteristics of training means used by track-and-field jumpers.

We have found out that in training exercises the take-off is performed using relatively independent BMs, similar to those recorded in competitive jumps. Being dependent of the motor task (conditions of performing the exercise), key biomechanisms appear to be interdependent on the dynamic level, i.e. the contribution of one of them affects that of the others. The role and contribution of the BMs depend on the type of an exercise or conditions of its execution, initial conditions, and a motor task set to an athlete. There exist different ways of realization of any BM as well as different interaction between BMs within the same jumping event.

Specific features of take-off in the examined exercises permitted to classify all the training means into four groups (**Table 8**):

- training means, mostly involving the BM of legs (take-off leg) and body extension (group I)

- training means, mostly involving the BM of swinging links (group II);

Exercise	BM of leg and body extension	BM of swinging links	BM of the "overturned pendulum"	Group
Long jump after a short approach run	17	68	15	I
Long jump over a hurdle	15	65	20	III
Long jump from a raised take-off platform (0.05 m)	13	65	22	II
Long jump from a lowered take-off platform (0.05 m)	10	74	16	III
Jump up with touching a suspended object by hand	13	70	17	IV
pattern of 3 hops	18	69	11	I

Table 8.
Contribution of different BMs in track-and-field technique development exercises (%).

• training means, mostly involving the BM of an "overturned pendulum" (group III);

• training means, involving the combination of the BM of swinging links with the BM of the "overturned pendulum" (group IV).

Thus, different special exercises are intended to exert specific effects on the structure of take-off, those effects being dependent on the specifics of the content and form of an exercise.

This comparison of technical drills differs from the conventional one, in which every kinematic or dynamic parameter of an exercise is compared with the similar parameter of an actual competitive jump.

Several specific exercises that are currently used in training athletes in different jumping events are listed below as examples (**Table 9**). All the exercises are classified into groups I – IV and may be recommended for practical use by athletes of a corresponding specialization. The list of exercises is not full because of the scope limitations for materials to be presented, but it provides general notion about aspects of training means selection for solving concrete training tasks taking into account jumpers' specialization.

Group	Jump event	Exercises
I	Triple jump Long jump Pole vault	After 4–6 running strides jump onto a pile of mats landing on a swinging leg. Important is to bring pelvis forward at the take-off.
III	High jump Long jump	After 4–6, running strides jump in a "stride" over 2 hurdles; the distance between the take-off spot and the first hurdle is from 180 to 220 cm, the distance between the hurdles 80–90 cm.
III	Triple jump Long jump Pole vault	After 6–7 strides of a direct approach, run make a long jump over a bar set at the height 70–80, 90–100, 110–120, or 130–140 cm, taking off in 80–90 cm from the nearest upright.
IV	High jump Long jump	After 4–8 running strides make a long jump attempting to touch a suspended object by the chest or head. Take off in a distance of 2–2.5 m from the projection of the hanging object.
II	Triple jump Long jump	Alternate leg bounds in a pattern: floor (the take-off leg) – floor board (the swinging leg) – low vaulting horse (the take-off leg) with touching a suspended object.

Group	Jump event	Exercises
I	Triple jump Long jump High jump	A barbell 20–30 kg on the shoulders. After 2–3 walk steps plant the take-off leg on the board (5–10 cm lower than the surface of the approach) and pushing off by the swinging leg quickly straighten the take-off leg with the following swing by the swinging leg.
II	High jump Long jump	Standing on a gymnastic bench, feet shoulder width apart, a barbell (10–30 kg) on the shoulders. Step forward, straighten the support leg bringing the swinging leg forward simultaneously, and putting it on a bar of wall bars. The distance between the bench and the wall bars – 200 – 280 cm.
II	Triple jump Long jump Pole vault	Stand with one foot in front, the other one behind the trunk. Grip the hanging rope at the head level. Push off the take-off leg bringing pelvis to the rope. While moving the rope in a circle, perform giant strides. Important is to bring pelvis forward in proper time.
IV	Triple jump Long jump Pole vault	Stand on a platform 70–80 cm high, the swinging (take-off) leg in front, the take-off (swinging) leg behind. Jump down landing on the take-off (swinging) leg and bounce up onto a platform 20–30 cm high trying to shorten the touch-down to take-off phase as much as possible. The take-off is similar to that of a long (high) jump.
IV	High jump	After 6–8 approach strides in a curve leaning into the arc, make a long jump attempting to touch a suspended object by the chest or head. Take off in a distance of 2–2.5 m from the projection of the hanging object.
IV	High jump	The same exercise as the previous one, but after a swing turn the knee of the swinging leg inside (toward the take-off leg). When taking off turn the trunk so that the shoulder opposite to the take-off leg is brought forward.
III	Triple jump Long jump High jump	Jump down from a platform 50–80 cm high, land on one foot, and jump over a hurdle trying to make the touch-down to take-off phase as short as possible. The height of the hurdle should be gradually increased. The distance from the platform to the hurdle 3–5 m.
III	Triple jump Long jump High jump	The same as the previous one, but landing on both feet.
IV	Triple jump Long jump	After 2–4 running strides make a triple jump starting from a raised platform. In the third phase (the jump) clear a hurdle set at 50–80 cm height.
I	Pole vault	Starting position (s. p.): Grip onto a pole (in a vertical position) by the right hand, keeping the arm straight, do one step backward and grip the pole by the left hand 30–40 cm lower than the right one. Without moving in any direction tries to touch the pole by chest lifting the knee of the swinging leg.
I	Pole vault	S. p. – similar to the previous one. While stepping forward, do a take-off and hang on to the pole.
II	Pole vault	After 4–6 running strides take off and hang on the pole. At the end of the legs swing make a half turn counterclockwise to face the approach runway.
I	Pole vault	After an approach run jump and grip on to a hanging rope, swing legs upward–forward and turns counterclockwise. Clearing a bar can be added.
II	Pole vault	When performing a pole vault swing legs upward trying to reach the upper end of the pole by them.

Table 9.
Specific exercises that are currently used in training athletes in different jumping events.

Abbreviations

BM   Biomechanism
CFJ   coxofemoral joint
KJ   knee joint

AJ   ankle joint
COG   center of gravity
CNS   central nervous system
SP   starting position
MTU   muscle-tendon unit

Author details

Mikhail Shestakov* and Anna Zubkova
Federal Science Center of Physical Culture and Sport (VNIIFK), Moscow, Russia

*Address all correspondence to: mshtv@mail.ru

IntechOpen

References

[1] Delis I, Hilt PM, Pozzo T, et al.. Deciphering the functional role of spatial and temporal muscle synergies in whole-body movements. Scientific Reports. 2018;**8**:8391

[2] Hay JG. The Biomechanics of Sports Technique. 4th ed. Englewood Cliffs, N.J: Prentice Hall; 2002

[3] Lazarević M. Mechanics of human locomotor system. FME Transactions. 2006;**34**:105-114

[4] Pandy MG, Berme N. A numerical method for simulating the dynamics of human walking. Journal of Biomechanics. 1988;**21**(12):1043-1051

[5] Seluyanov VN, Shalmanov An A, et al. Biomechanisms as the basis of biomechanics of human movements in sport. Theory and Practice of Physical Education. 1995;7:6-10

[6] Shestakov M. Use of Modeling in conducting technical training. Theory and Practice of Physical Education. 1998;3:51-54

[7] Shestakov M, Zubkova A. Peculiarities of Muscle-Tendon Mechanics and Energetics in Elite Athletes in Various Sports. In: Contemporary Advances in Sports Science. London: IntechOpen; 2021. pp. 159-176

[8] Kawato M. Internal models for motor control and trajectory planning. Current Opinion in Neurobiology. 1999;**9**(6):718-727

[9] Wolpert DM, Kawato M. Multiple paired forward and inverse models for motor control. Neural Networks. 1998;**11**:1317-1329

[10] Friston K. What is optimal about motor control? Neuron, 72(3), 488-498. Trends in Cognitive Sciences. 2011;**13**(7):293-301

[11] An S, Al S. Basic Mechanisms of Contacts with a Surface in Jumps. Moscow: GCOLIFK; 1990. p. 48

[12] Vaverka F, Jandačka D, Zahradník D, et al.. Effect of an arm swing on countermovement vertical jump performance in elite volleyball players: FINAL. Journal of Human Kinetics. 2016;**53**:41-50

[13] Linthorne NP. Analysis of standing vertical jumps using a force platform. American Journal of Physics. 2001;**69**:1198-1204

[14] Panoutsakopoulos V, Kotzamanidou MC, Papaiakovou G, Kollias IA. The ankle joint range of motion and its effect on squat jump performance with and without arm swing in adolescent female volleyball players. Journal of Functional Morphology and Kinesiology. 2021;**6**:14

[15] Fujibayashi N, Otsuka M, Yoshioka S, Isaka T. Technical strategy of triple jump: Differences of inverted pendulum model between hop-dominated and balance techniques. The Journal of Sports Medicine and Physical Fitness. 2018;**58**(12):1741-1751

[16] Kuo AD. The six determinants of gait and the inverted pendulum analogy: A dynamic walking perspective. Human Movement Science. 2007;**26**:617-656

[17] Aeles J, Lichtwark G, Peeters D, Delecluse C, Jonkers I, Vanwanseele B. Effect of a prehop on the muscle-tendon interaction during vertical jumps. Journal of Applied Physiology (Bethesda, MD: 1985). 2018;**124**(5): 1203-1211

[18] Prilutsky BI, Zatsiorsky VM. Tendon action of two-joint muscles: Transfer of mechanical energy between joints during jumping, landing, and running. Journal of Biomechanics. 1994;**27**(1):25-34

[19] Umberger BR. CSCS mechanics of
the vertical jump and two-joint muscles.
Strength and Conditioning.
1998;**20**(5):70-74

[20] Bobbert MF, van Ingen Schenau GJ.
Coordination in vertical jumping.
Journal of Biomechanics.
1988;**21**(3):249-262

[21] Iida F, Rummel J, Seyfarth A.
Bipedal walking and running with
spring-like biarticular muscles. Journal
of Biomechanics. 2008;**41**(3):656-667

[22] Aerts P. Vertical jumping in Galago
senegalensis: The quest for an obligate
mechanical power amplifier.
Philosophical Transactions: Biological
Sciences. 1998;**353**(1375):1607-1620

[23] Astley HC, Roberts TJ. Evidence for
a vertebrate catapult: Elastic energy
storage in the plantaris tendon during
frog jumping. Biology Letters.
2012;**8**(3):386-389

[24] Hill AV. The heat of shortening and
the dynamic constants of muscle.
Proceedings of the Royal Society of
London. 1938;**B126**:136-195

Chapter 8

From Exercise Physiology to Network Physiology of Exercise

Natàlia Balagué, Sergi Garcia-Retortillo,
Robert Hristovski and Plamen Ch. Ivanov

Abstract

Exercise physiology (EP) and its main research directions, strongly influenced by reductionism from its origins, have progressively evolved toward Biochemistry, Molecular Biology, Genetics, and OMICS technologies. Although these technologies may be based on dynamic approaches, the dominant research methodology in EP, and recent specialties such as Molecular Exercise Physiology and Integrative Exercise Physiology, keep focused on non-dynamical bottom-up statistical inference techniques. Inspired by the new field of Network Physiology and Complex Systems Science, Network Physiology of Exercise emerges to transform the theoretical assumptions, the research program, and the practical applications of EP, with relevant consequences on health status, exercise, and sport performance. Through an interdisciplinary work with diverse disciplines such as bioinformatics, data science, applied mathematics, statistical physics, complex systems science, and nonlinear dynamics, Network Physiology of Exercise focuses the research efforts on improving the understanding of different exercise-related phenomena studying the nested dynamics of the vertical and horizontal physiological network interactions. After reviewing the EP evolution during the last decades and discussing their main theoretical and methodological limitations from the lens of Complex Networks Science, we explain the potential impact of the emerging field of Network Physiology of Exercise and the most relevant data analysis techniques and evaluation tools used until now.

Keywords: complex systems, circular causality, nonlinear dynamics, timescales, self-organization

1. Introduction

Exercise physiology (EP), the study of how the body adapts physiologically to the acute and chronic stress of exercise or physical activity, has evolved extensively since the beginning of the early twentieth century. Due to an increased interest in exercise and health, initially motivated by the poor physical capacity of soldiers, is today a scientifically founded branch that provides the basis of physical fitness, exercise performance, training, testing, and rehabilitation programs addressed to all types of population, including elite athletes and clinical patients. Its potential to enrich Basic Physiology and diverse fields such as Sports Medicine, Sports Rehabilitation, Sport Science, or Training Science is still undervalued and has to be rediscovered under the framework of Complex Systems and Network Science approaches [1].

IntechOpen

In this chapter, the present and future of EP will be overseen from a historical and scientific perspective. The main limitations of the EP available evidence-based research, strongly influenced by excessively simplified theoretical and methodological assumptions, will be discussed using the example of the exercise-induced fatigue. Finally, the research approach of the new emerging field of Network Physiology of Exercise, focused on the coordination and integration among physiological systems across spatiotemporal scales (from the subcellular level to the entire organism), will be presented.

2. Evolution of Exercise Physiology

It is essential to understand the EP history when approaching the future [2]. As any scientific branch, the EP evolution has been constrained by multiple and multilevel factors acting at different timescales such as financial possibilities, organizational and ideological positions. Although historical data pertaining to EP spans more than 2000 years, first research contributions correspond to the early twentieth century, which was characterized by an increasing specialization and sub-specialization in many scientific fields. This state of affairs brought about a flood of fragmentation in science that promoted the naissance and development of the main EP research labs in the world.

First works, initiated by Scandinavian scientists, were related to metabolism and heat production during exercise and recovery. Maximal oxygen uptake was described as the upper limit of performance [3], and lactate production (from glucose metabolism) was related to fatigue [4]. Research was focused later on circulation, muscle physiology, or environmental physiology and provided the basis of exercise as medicine. While the major concern of research after the World War II was the health and fitness of soldiers, the most recent concern is the obesity epidemic and other diseases related to the food abundance and lack of physical activity.

With the development of labs and the creation of world organizations such as the American College of Sports Medicine (ACSM), and the European College of Sport Sciences (ECSS), the field of EP has become enormously specialized in the last years, and EP researchers usually work in one area (e.g., cardiovascular, muscular, etc.). This has produced a loss of the original essence of Physiology, the unique branch of Biology specifically dealing with synthesis and integration. Although technological advances have led to create more sophisticated and better equipped labs, the type of inquiry and research focus of EP has been kept in general quite immutable, and clinical exercise physiologists keep mainly directed to testing energy production (e.g., aerobic power, anaerobic thresholds, etc.) [5].

Influenced by reductionism, Biology has traditionally emphasized: the decomposition of systems responsible for a given phenomenon into component parts and processes. Identifying such components and describing their mechanisms apart during exercise have been one of the main EP endeavors. Using a range of experimental models from cells to animals and humans, main approaches have laid the description of the biological mechanisms of temporary and persistent functional changes in response to acute and repeated exercise [5, 6]. It is worth noticing that despite the evolution of biology and the technological advances of the last years, the initial theoretical assumptions of EP have kept almost intact.

Even if "why" questions, related to teleological explanations, have been traditionally avoided because the final purpose of physiological systems is assumed to be unknown or nonexistent, research questions often reflect the excessively simplistic assumptions that have characterized EP from the very beginning (see Section 3 for further detailed explanation). The linear and reductionist approach of the scientific production is reflected in the redundant expression "Effects of," highlighted in the content analysis done on over 22.000 ECSS abstracts submitted during almost two decades [7].

This expression reflects a very specific mode of inquiry, and its consequential data acquisition tools, analysis techniques, and inductive methods, commonly and uncontroversially used in EP research.

A century of reductionist research has produced a lot of information and descriptive knowledge, some obtained through very well-designed experiments, but has provided only a partial understanding of exercise-related phenomena, and led to several controversial findings. In fact, some of the main questions still remain without clear responses. For instance, which are the limits of performance, what limits VO_2max, what causes fatigue, why are there responders and non-responders to exercise, etc. The extant controversies seem strongly affected by the excessively simplified theoretical and methodological assumptions that characterize the field.

2.1 Molecular Exercise Physiology. Are explanations only in the cell?

Due to the lack of clear responses to the main topics obtained investigating at system and organ levels (e.g., cardiovascular system, muscle), the reductionist rationale led EP investigation, and medical investigation in general, in which disease is increasingly understood in molecular terms, toward microscopic levels (Molecular Biology, Genetics, and OMICS technologies). The acknowledgment of the role of exercise on health status and the pathway toward personalized medicine has also reinforced the micro-level research focus of EP [8–11].

While the 1970s were the decade of Biochemistry, the 1980s represented the Molecular Biology era. Technological advances played a fundamental role in this evolution. The introduction of DNA microarrays, a fast technology to study thousands of DNA and protein molecules simultaneously, supposed a revolution in biological research. Coupled with computational methods, pushed the development of Systems Biology (e.g., [12]), a branch that focuses on complex interactions within biological systems, and enabled to investigate the behavior of the genes of an organism under different conditions [13]. The identification of new biomarkers, the improved sensitivity and specificity of the existing ones, and new insights into the personalized therapeutic strategies to improve athletic performance and human health through precision exercise medicine is the main aim [14].

Research in Molecular Exercise Physiology and "sportomics" [15–17] is mostly focused on omics data collection and analysis efforts to catalog exercise-regulated pathways. Although Molecular Biology dwells on dynamic principles, the dominant research methodology in Molecular Exercise Physiology, and Integrative Physiology [10, 18] keep focused on non-dynamic bottom-up group-pooled statistical inference modes of inquiry. Main properties of CAS as synergies, established not only horizontally (e.g., among molecules) but vertically (among molecular, cellular, tissular, organ, and system levels), are neglected. The embeddedness of lower levels in upper levels, the circular causality (bottom-up, top-down) relationship among the levels, the different timescales of activity, the nonlinear dynamic processes that suffer qualitative changes through self-organization are also some of the main neglected properties.

Many component processes can lose or gain on significance during exercise; for instance, as fatigue develops (see Section 2.2). Physiological and psychobiological synergies compensate critical quantitative values registered at micro-level keeping stable behavioral variables registered at action level. In biological systems, the same effect at microscopic level can be produced by many different macroscopic phenomena. In addition, genes are dominantly pleiotropic, that is, the same gene can be involved in different physiological effects and states. Hence, the informativeness of the microscopic (Molecular Biology levels) may be at best an initial point in a much more elaborate study of the organism-environment interaction to conclude on the real macroscopic phenomenon that is involved in a health or performance problem.

Personal health and performance cannot be reduced to molecular and genetic levels as medicine cannot be geneticized. In addition, the network relations do not operate only bottom-up, but also top-down, that is, from the entire person to the genes level following the circular causality property of complex systems [19].

The view that everything can be explained at microscopic (molecular) level directly implies that health can be intervened and repaired only at this level, that is, through pharmacological substances. However, a person-environment approach [19] implies that health and performance are products of the interaction of many different levels, and health can be also improved intervening at environment or psychological level. In fact, interventions at macroscopic (social, psychological) level have been proven as crucial in changing the processes at subcellular level [20, 21]. Results seems to show that in a sick society, where more often than not, the competition for socially imposed success becomes a goal for itself, there is an increased likelihood of cell aging and poor personal health [22]. In multilevel complex networks, the macroscopic ambience strongly constrains its embedded components [23, 24]. While the use of pharmacological substances may be promoted somewhat by the big financial benefits that lie behind, mental interventions, usually cheaper and requiring only the development of self-knowledge and self-discipline [20], receive in general less scientific attention. In particular, exercise is a privileged type of intervention because it may affect all personal levels in a correlated cascade way [7, 25].

Systems Biology and Integrative Physiology strive for the same goal: to understand Biology whole-istically [18]. Systems Biology is focused on systems operating at a cellular level and has evolved over the past decade (called the "omics era") as a direct result of advances in high-throughput molecular biology platforms and associated bioinformatics. In fact, Systems Biology has been described as: "the study of an organism, viewed as an integrated and interacting network of genes, proteins and biochemical reactions which give rise to life." Instead of analyzing individual components or aspects of the organism, such as sugar metabolism or a cell nucleus, systems biologists focus on all the components and the interactions among them, all as part of one system. These interactions are ultimately responsible for an organism's form and function [26].

Physical exercise dictates the magnitude and pattern of how networks of genes, proteins, and biochemical reactions will integrate and interact. This is an important point, because to the exercise physiologist most, if not all, cellular-network-based change will be secondary to the physiological stimulus causing that change, e.g., muscle contraction, rather than originating at the level of the network per se. This in essence is one important difference between the molecular biology focus at the core of Systems Biology and functional feedback approach of Integrative Physiology, a difference eloquently described by Noble [27]. In conclusion, although there is a tremendous potential for omics approaches to fill critical gaps in our understanding of the integrative networks underlying the health benefits of exercise [28] and allow going beyond the one-size-fits-all model of prescription [29], the complexity and interconnectedness of exercise biological networks cannot be unraveled and understood by studying single tissues or molecular targets alone. They require dynamic, multilevel, and global approaches. For instance, whether molecular processes can inform about the level of stress, the macroscopic phenomenon of stress cannot be explained only at microscopic level.

2.2 From microscopic to systemic hypotheses in Exercise Physiology. Example of exercise-induced fatigue research

The research performed to respond to the question about the causes of exercise-induced fatigue and spontaneous task failure is a good example of the current tendencies in EP research.

Over the last century, physiologists have tried to find the etiology and underlying mechanisms of exercise-induced fatigue [30]. Despite a wealth of knowledge about individual components intervening in the fatigue process and their adaptation to different types of exercise, they have failed to detect a single component or process responsible of the phenomenon and the limits of exercise tolerance in general [31]. The questions of "what causes exercise-induced fatigue" or "are limiting mechanisms central or peripheral, are there in the brain or in the muscle [32, 33]?" are clear examples of the type of inquiry searching for cause-effect relationships and the fragmentation tendencies derived from the reductionist models applied to the EP research.

The research investigating central and peripheral mechanisms of exercise-induced fatigue has not provided either a clear response to the question [34–37]. The impaired action potential propagation, the inhibition of reflex mechanisms, the stimulation of chemical and nociceptive afferent signals, the corticospinal stimulation changes, the increase in extracellular serotonin, the cytokines liberation, the muscle acidosis, the accumulation of NH_4, $H+$, Mg_2+, Pi, the hyperthermia, the inhibition of Ca_2+ liberation, the glycogen reduction, the increase of $K+$ and free radicals are all associated processes to the fatigue development but cannot explain it.

In a similar way, assumed cause-effect or dose–response relationships among biochemical and performance variables have been proven to be often wrong. For instance, lactic acid, initially thought to be the consequence of oxygen lack in contracting skeletal muscle and related to the limits of high-intensity short-duration exercise, now it is recognized as being formed under fully aerobic conditions and associated to ergogenic and antifatigue properties [38, 39].

Fatigue is a macroscopic phenomenon and reflects itself in macroscopic behavior of performers [40]. Microscopic processes associated to it are not linearly independent (as it is usually tacitly assumed), and their total effects cannot be treated as a sum of individual effects. In particular, when knowing that there is a circular causality spread over the levels in all complex systems. Instead of focusing on isolated central and peripheral processes, the exercise-induced fatigue and task failure can be studied at behavioral coordination level, which integrates all network levels [41]. Using a macroscopic kinematic variable extracted at action level, the authors studied the time-variability properties of the elbow angle, considered as order parameter, during an effort performed until exhaustion. Critical behavior such as critical slowing down, enhancement of fluctuations, and correlation enhancement in interlimb coordination was reproducibly observed [42]. In this way, fatigue was understood as a process that leads to a *system-level* phase transition (spontaneous task disengagement), due to the circular causality mechanism that spreads over the levels in CAS.

In this way, it was hypothesized that the spontaneous task failure consists of a percolation process, produced by the impaired ability of the psychobiological network system to make the necessary short-term adjustments for negotiating the imposed external workload. In this scenario, the spontaneous task failure/disengagement, represent a giant (at systemic level), protective inhibitory fluctuation that causes a temporary abrupt switch to a lower energy expenditure level, a critical phenomenon prominent in complex systems, such as human psychobiological networks. This means that the loss of stability at systemic level is the cause of the task disengagement and not a singular process at singular level that can be pinpointed.

3. Limitations of Exercise Physiology. Contrasting approaches from complex network science

In contrast to decomposition, EP has paid much less attention to re-composition of physiological mechanisms [43]. The greatest challenge today, not just in Biology but in all of science, is then to reassemble such decomposed mechanisms capturing the key properties of the entire ensembles [44]. If composing and closing the sequence of the Krebs cycle and/or describing the sequence of reactions of the glycolysis was years ago a key innovation, the challenge today is defining the nonlinear dynamics of embedded network processes under constraints.

New technologies and interdisciplinary work have promoted the introduction of complex systems thinking on EP, but there is still a long way to go. The sophistication and capacity of modern technology, able to shift the landscape of basic life sciences research from that of traditional biological reductionism to a much more integrative, holistic systems approach [45], are not enough. In fact, a change from a reductionist to holistic paradigm cannot be achieved only via the technical world. Together with new techniques and technologies, the development of new theoretical assumptions, conceptual frameworks, and analysis tools is necessary. New ways of doing and understanding based on complex systems, proven as successful in other disciplines, should be also implemented in EP [46]. As pointed by Greenhaff and Hargreaves [26], "perhaps the tools of Systems Biology should be viewed increasingly as a valuable addition to the arsenal that exercise scientists can use to interrogate physiological function and adaptation" (**Table 1**).

a. Theoretical assumptions

Instead of fragmenting and studying separately the functions of different physiological components and processes, the focus of Networks Science is put on the interaction dynamics among such components and processes. Classical cybernetics, inspiring the basic biological control system model of EP, is replaced by Dynamical Systems Theory (DST), which provides concepts and tools to describe

Theoretical assumptions	Exercise Physiology	Complex Networks Science
Systems	Dominated by their components	Dominated by the interactions among components
Theory	Cybernetic Control Theory	Dynamic Systems Theory
Control	CNS as the main regulator/programmer	Parametrically regulated system
Mechanisms	Homeostatic	Homeodynamic
Methodological traits	**Exercise Physiology**	**Networks Science**
Variables	Isolated variables	Networked collective variables
Measures	Means and max values of variables	Connectivity/Synergy/Coordination dynamics
Data acquisition	Group-pooled data	Intra-individual time series
Analysis	Population to individual generalization	Individual to population generalization
Relations	Bottom-up (from micro to macro levels) statistical inferences	Bottom-up and top-down (circular causality) multilevel dynamic interactions

Table 1.
Contrast of some limiting theoretical and methodological assumptions of EP research with assumptions based on Networks Science.

and study the coordinative changes occurring in the physiological network over time.

According to classical cybernetics, different components and processes operate through feedback loops to maintain physical or chemical physiological parameters constant (homeostasis). The predictions of this "engineering" approach are linear, i.e., proportional between inputs and outputs and are displayed through descriptive block diagrams, commonly used in EP to represent how organic structures and processes interact. The basic assumption of these diagrams is that of time-invariant encapsulated modules, processes, and regulation profiles. While the concept of feedback works fine in simple systems that have only two parts to be joined, each of which affects the other, when a few more parts are interlaced together, the system very quickly becomes impossible to treat in terms of explicit feedback circuits.

In complex systems, there is no reference state with which feedback can be compared and no place where comparison operations are performed. Nonequilibrium steady states emerge from the nonlinear interactions among the system's components, but there are no feedback-regulated set points or reference values as in a thermostat. For instance, it is not possible to explain through feedback loops phenomena such as the fatigue-induced task disengagement [47], the overtraining syndrome [47, 48], or the macroscopic emergence of noncontact injuries [49].

Feedback homeostatic mechanisms are replaced from a Networks Science point of view by the concept of homeodynamics or dynamic stability, i.e., a constantly changing interrelatedness of body components and processes while an overall equilibrium is maintained [50].

It is common among exercise physiologists to propose conceptual models where the main regulator or programmer is the Central Nervous System (CNS) (see, e.g., [45, 51]). Integrative Physiology also neglects that CAS does not need any internal or external programmer to regulate their functions [52]. Properties of such functions (i.e., stability, instability, switches among states, etc.) are parametrically regulated, and the CNS is also a regulated subsystem. This means that physiological states emerge from the interaction among multilevel system components (the CNS being another component) through a self-organized process. The search for the ultimate high-level regulator would end in infinite regress (who regulates the regulator that regulates...?) and represents a loan on understanding exercise-related physiological phenomena.

b. Methodological traits

Instead of isolated variables, the use of macroscopic collective variables is proposed because they behave as order parameters integrating all network levels and capturing the system organization. The dynamics of such variables, reflected in their time variability properties, may inform about the interactions among system components and may help detecting different states and anticipate qualitative changes.

Instead of using only molecular data to establish bottom-up statistical timeless inferences from micro to macroscopic phenomena, the study of the time-variability properties of behavioral macroscopic variables, extracted at action level (e.g., the elbow angle in Section 2.2) during exercise, may inform about the vicinity of qualitative changes. This approach helps to detect dynamic features such as stable and unstable states, critical behavior (phase transitions, bifurcations, critical slowing down, enhancement of fluctuations). It is worth to remark that such behavior can be produced at different quantitative values of physiological parameters or set points [41].

Specific proposals of Network Physiology are developed in sections 4 and 5.

Most of the available research on EP, based on inductive analytical research, infers intra-individual phenomena from the analysis of inter-individual variations obtained through group data means and comparison designs. This approach has some basic debatable methodological issues that should be discussed.

The main aim of physiological (systemic, biochemical, genetic, epigenetic) research is to find the mechanisms of regulation and causal changes that occur at intra-individual (i.e., organism) level as an effect of various internal and external factors. In other words, the intra-organismic processes and not the population are the explanatory target. It is important to note here that intra-individual variability and co-variability unfold in time and hence, need to be measured through time series analytical tools. While the problem of sample to population generalization has been much discussed, investigated, and used, in inferential statistics, much less attention has been focused on the question of sample or population to individual generalization. A tacit assumption has been that the results obtained at sample and generalized to population level are representative of the changes of a "typical" (i.e., average) individual [53–55]. In other words, the group-pooled data merely would enhance the typical phenomenon that already exists in every and each individual. However, in order for this assumption to be correct, there are some strict conditions to be fulfilled. These conditions are the non-violation of ergodicity[1] assumption. On the other hand, pooling over group subjects is the predominant research practice in exercise and health-related research. Even the state-of-the-art analytical software packages for time series analysis [56] are based on pooling-over-subjects approaches.

The correctness of the tacit assumption of ergodicity conditions, to our knowledge, has never been explicitly tested EP. Hence, the generalization of results from population to individual (or between clusters of individuals) may be typically not valid for developmental biological systems. This means that the structure of causal changes may drastically vary from individual to individual, and these differences are not detectable at the group-pooled level. Some approaches to overcome these serious problems have been proposed [57–60].

Using molecular data, Integrative Exercise Physiology and OMICS techniques are focused on establishing non-dynamical bottom-up statistical inferences between micro- and macro-level states, ignoring one of the main properties of CAS: the tendency to form multilevel synergies. Synergies acting at different levels (e.g., molecular, cellular, tissue, organ, etc.) allow reciprocal compensations among physiological components and processes to satisfy a task goal during exercise. Such synergies are flexible assembled patterns of coordination, which form emergent structures and functions responding to the exercise requisites. Without them, life would not, and could not, exist. Through circular causality relations, components form new synergies, which govern, in turn, the components' behavior [61, 62]. While computer scientists build programs that tell circuits what to do, nature builds synergies [63, 64].

Synergetic is also manifested through the CAS property of degeneracy: different components produce the same function, and different synergies may be activated to attain the same task goal [65, 66]. For instance, different motor units cooperate and reciprocally compensate their activation over several timescales to perform an effective or functional motor action over time during a running competition. The self-assembled, adaptive interactions of CAS underpin also another robustness enabling property: pleiotropy or multifunctionality, that is, the same components

---

[1] Ergodicity: The system's stochastic evolution in time is stationary (stationarity assumption) and the structure of the intraindividual multivariable dynamics is the same in all individuals (the homogeneity assumption). Typically, jointly these conditions are not fulfilled in biological systems.

may be assembled to produce multiple functions. For instance, the skeletal muscle, with genuine/primordial contractile functions, may exert as well immunological and endocrine functions [67, 68]. Such properties enable CAS to switch between diverse coordinative states and maintain a metastable dynamic [69].

4. Network physiology of exercise: A paradigm shift

Dynamic models, initially rejected by biologists, initiated some 40 years ago a paradigm shift in general biology [70, 71], molecular and cell biology [72], genomics [73], and all the "omic"-based approaches [74], which are now at the forefront of science. Such interaction-based approaches have started to spread in EP and relevant fields of medical research such as cancer [75]. However, Physiology, and in particular EP, should do a substantial effort for reassembling biological processes and focusing not only on horizontal interactions at molecular level and establishing non-dynamical statistical inferences to the entire person (e.g., performance or health status, see **Figure 1**, left), but integrate all vertical network levels (e.g., molecular, cellular, tissue, organ, systems, see **Figure 1**, right). That is, avoiding the gap between micro and macro structures and functions and considering the multiple vertical synergies that may act among them.

Network Physiology of Exercise (NPE) emerged inspired by the field of Network Physiology [76–84] and Networks Science [1]. Network Physiology addresses the fundamental question of how physiological systems and subsystems coordinate, synchronize, and integrate their dynamics to optimize functions at the organism level and to maintain health. It aims at uncovering the biological dynamic mechanisms [85–88] since it satisfies both the mechanistic requirement of structure and localization (e.g., nodes and edges/links in dynamic networks may represent localized integrated organ systems, subsystems, localized components or processes, and interactions among them across various levels in the human organism) and the requirement of dynamical invariance and generality that is enabled by dynamical systems approach [89].

In the context of exercise, NPE aims to transform the theoretical assumptions, the research program, and the current practical issues of current EP. It focuses the research efforts on improving the knowledge of the nested dynamics of the vertical (among levels) and horizontal (among organs and components) network

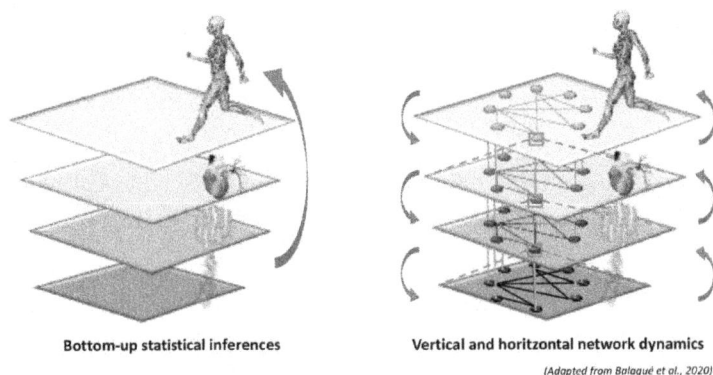

Bottom-up statistical inferences   Vertical and horitzontal network dynamics

(Adapted from Balagué et al., 2020)

Figure 1.
Contrast between Molecular Exercise Physiology (left), focused on non-dynamical bottom-up statistical inference techniques, and Network Physiology of Exercise (right), focused on the nested dynamics of the vertical and horizontal physiological network interactions.

interactions to understand how physiological states and functions emerge under different constraints and contexts.

Studying the organism as a dynamical system means studying a set of variables that interact over time, that is, their time series, that may exhibit various patterns. DST comprises a highly general set of mathematical concepts and techniques for modeling, analyzing, and interpreting these patterns in time series data. Therefore, DST is not applied exclusively to the area of biomedical sciences, it can also be used to describe social and psychological phenomena, among others [90–92].

Many physiological mechanisms exhibit oscillations or more complex dynamical behavior, which is crucial for orchestrating operations within the mechanism. Such complex behavior is non-sequential, because some of the interactions in the mechanism are nonlinear, and the system is open to energy. Initial positive adaptations of physiological functions are followed by stagnation or decrease of such functions when workload increases further (e.g. overtraining syndrome, see [47, 48, 93]).

Interactions, generate novel information that determines the future of elements, and thus of the system itself [94]. The interaction-dominant dynamics of humans, in contrast with the typical component-dominant dynamics of machines [95], has been emphasized in the EP literature [41, 96, 97]. This means that the behavior of CAS cannot be simply explained through linearly independent variability sources, processes, or local mechanisms. For instance, exercise physiologists cannot rely on critical quantitative endpoints in cardiovascular, respiratory, metabolic, or neuromuscular systems to explain the limits of performance [31, 98, 99] and should reformulate their research hypothesis on the basis of CAS properties.

4.1 Network Physiology of Exercise. Data analysis techniques

Novel data analysis techniques have been successfully applied in the context of Network Physiology to explore how physiological systems dynamically integrate as a network to produce distinct physiologic functions [80, 85, 86, 100]. The goal of such tools is to develop a general theoretical framework and a computational instrumentarium tailored to infer and quantify interactions among diverse dynamical systems—specifically, (i) systems of oscillatory, stochastic, or mixed type; (ii) systems with noisy, nonstationary, and nonlinear output signals; (iii) systems acting on widely different timescales from milliseconds to hours; (iv) systems coupled through multiple coexisting forms of interaction. Some of the most relevant data analysis techniques to infer couplings among several physiological systems, with potential to be utilized under exercise settings, are the following:

4.1.1 Time delay stability (TDS) method

Integrated physiologic systems are coupled by feedback and/or feed-forward loops with a broad range of time delays. To probe the network of physiologic coupling, a novel concept has been introduced, time delay stability, and a new TDS method has been developed to study the time delay with which modulations/bursts in the output dynamics of a given system are consistently followed by corresponding modulations in the signal output of other systems. Periods with constant time delay indicate stable interactions, and stronger coupling between systems results in longer periods of TDS (**Figure** 2) [80, 101]. Thus, the strength of the links in the physiologic network is determined by the percentage of time when TDS is observed: higher percentage of TDS corresponds to stronger links. To identify physiologically relevant interactions, represented as links in the physiologic network, we determine a significance threshold level for the TDS based on comparison with surrogate data:

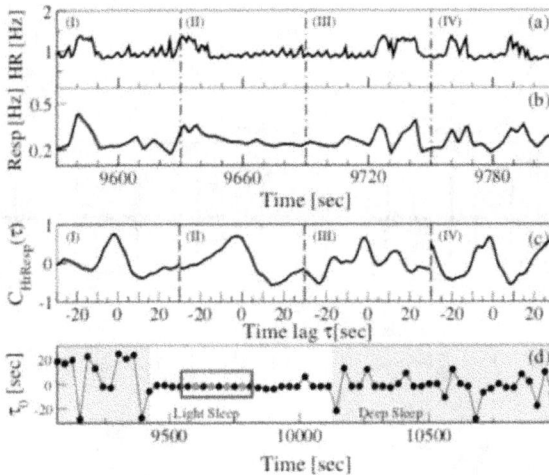

Figure 2.
Schematic presentation of the TDS method: Segments of (a) heart rate (HR) and (b) respiratory rate (Resp) in 60 sec time windows (I), (II), (III) and (IV). Synchronous bursts in HR and Resp lead to pronounced cross-correlation (c) within each time window in (a) and (b), and to a stable time delay characterized by segments of constant τ_0 as shown in (d)—four red dots high-lighted by a blue box in panel (d) represent the time delay for the 4 time windows. Note the transition from strongly fluctuating behavior in τ_0 to a stable time delay regime at the transition from deep sleep to light sleep at ~9400 sec and inversely from light sleep back to deep sleep at ~10,100 sec (shaded areas) in panel (d). The TDS analysis is performed on overlapping moving windows with a step of 30 sec. Long periods of constant τ_0 indicate strong TDS coupling.

only interactions characterized by TDS values above the significance threshold are considered. The TDS method is robust and can track in fine temporal detail how the network of connections between organ systems changes in time. The method is general and can be applied to diverse systems.

4.1.2 Phase synchrogram algorithm (PSA)

Nonlinear oscillatory systems are characterized by nonidentical eigenfrequencies and highly irregular signal output can synchronize even when their coupling is weak—i.e., their respective frequencies and phases "lock" at a particular ratio. Despite the significant difference in the periodicity of the cardiac and respiratory rhythms represented by the heartbeat and respiratory intervals, and despite the complex noisy variability in the cardiac and respiratory signals, previous work found that episodes of heartbeat-respiration phase synchronization emerge. Previous authors developed a synchrogram algorithm able to identify segments of cardiorespiratory phase synchronization and to track how the degree of this nonlinear form of coupling changes in time and across physiologic states ([85]; **Figure 3**). The PSA synchrogram algorithm is robust and can identify interrelations between output signals of nonlinear coupled systems even when these signals are not cross-correlated [86]. Thus, the PSA can quantify the degree of coupling between nonlinear systems when other conventional methods (such as cross-correlation or cross-coherence analysis) cannot.

4.1.3 Cross-correlation of instantaneous phases (CCIP) method

Due to the nonstationary trends embedded in physiologic signals, traditional cross-correlation and cross-coherence analyses fail to accurately quantify the interrelation between physiological systems. This approach based on the cross-correlation between the instantaneous phase increments of the output signals of nonlinear

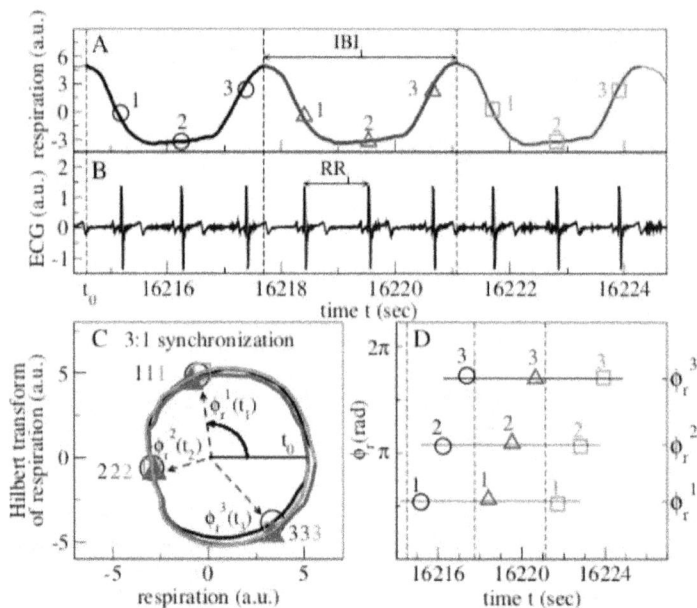

Figure 3.
Schematic presentation of the PSA method: (A) three consecutive breathing cycles and (B) a simultaneously recorded ECG signal. (C) Demonstration of phase synchronization between the heartbeats and respiratory cycles shown in (A) and (B). For each breathing cycle, all first heartbeats occur at the same respiratory phase $\varphi 1r(t)$, and all second and third heartbeats within each breathing cycle occur at $\varphi 2r$ (t) and $\varphi 3r$ (t), respectively (symbols collapse), indicating robust phase synchronization. (D) each heartbeat in the ECG signal (B) is shown with its phase φr (t) relative to the beginning of the breathing cycle in which it occurs. Different symbols represent heartbeats in different breathing cycles as in (A) and (C), and vertical dashed lines show the beginning of each breathing cycle. Three horizontal lines formed respectively by the first, second, and third heartbeats in the three breathing cycles indicate robust 3:1 phase synchronization despite noisy heart rate and respiratory variability.

coupled systems is not affected by the nonstationarity of the signals. Chen et al. [86] successfully applied this new approach to study cerebral autoregulation in healthy subjects and in stroke patients. The approach is sensitive to uncover previously unknown differences in the coupling between cerebral blood flow velocity and peripheral blood pressure in the limbs for healthy and post-stroke subjects (**Figure 4**). In contrast, linear cross-correlations and other traditional methods cannot identify changes in cerebral autoregulation after stroke.

4.1.4 Principal components analysis (PCA)

PCA has been recently conducted on the time series of several cardiorespiratory parameters during maximal exercise (**Figure 5**). The PCA pinpoints and quantifies whether the increment and decrement of time patterns from different physiological processes are statistically correlated. In this way, the magnitude to which time patterns of physiological responses covary in time is reflected. The covariation of several (two or more) cardiorespiratory parameters shows the mutual information that they share. This common variance, in turn, enables time patterns of single cardiorespiratory outcomes to be represented through fewer principal components (PCs). The PCs are obtained in decreasing order of importance and reflect the highest possible fraction of the variability from the original dataset. Thus, the total number of PCs indicates the level of coordination among the initial cardiovascular and respiratory parameters. More concretely, a dimensionality reduction is indicative of the creation of new

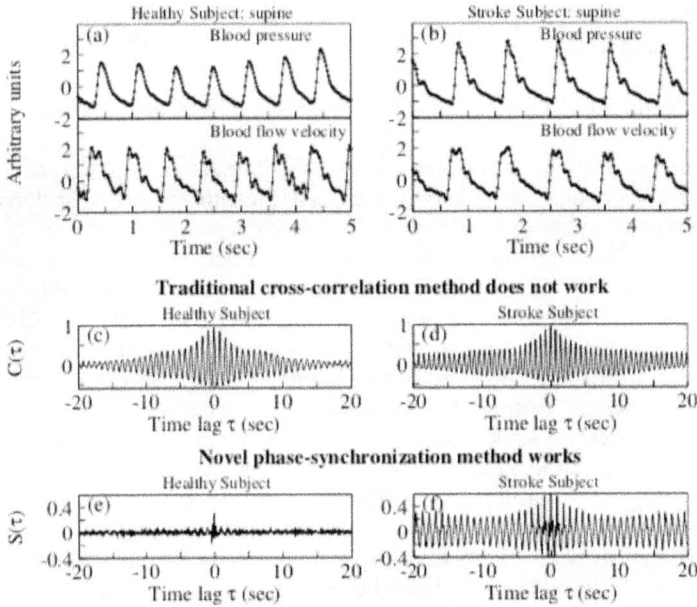

Figure 4.
The CCIP approach: Identifies breakdown of coupling mechanisms of cerebral autoregulation after stroke. Signals of peripheral blood pressure (BP) in the limbs and blood flow velocity (BFV) in the brain for (a) healthy and (b) post-stroke subject during a quasi-steady supine state without external perturbations. (c-d) Traditional cross-correlation function $C(\tau)$ for the BP and BFV signals for the same subjects shown in (a) and (b) does not identify differences in the BP-BFV coupling between healthy and post-stroke subjects. (e-f) In contrast, cross-synchronization function $S(\tau)$ obtained from the phase increments of BP and BFV signals using our CCIP method shows a significant difference in the BP-BFV coupling between healthy and post-stroke subjects, even though patterns in the pairs of BP-BFV signals are visually similar (a, b) for both healthy and stoke subjects.

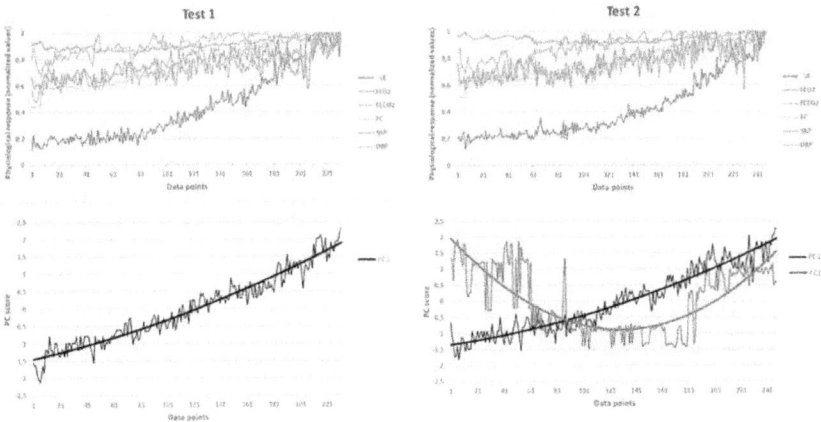

Figure 5.
Typical example of the reduction of cardiorespiratory variables to time series of cardiorespiratory coordination variables (PCs) in two consecutive maximal cardiorespiratory tests interspersed by 10-min resting: Test 1 and test 2. Top graphs: Original time series of the six selected cardiorespiratory variables in test 1 and test 2. Bottom graphs: Time series of PC scores (standardized z-values in the space spanned by PCs) in both tests. The six time series are collapsed to one time series (test 1) or two time series (test 2) as a consequence of the PC dimension reduction. The black and the red lines show the average trend of both processes as calculated by weighted least squares method. Data points of the x-axis of both graphs refer to the number of measurements recorded along the cardiorespiratory test.

coordinative patterns [102]; therefore, the reduction in the quantity of PCs suggests an enhancement in the efficiency of cardiorespiratory system [103].

4.2 Evaluation tools based on Network Physiology: Network-based biomarkers

The common testing variables used in Exercise Physiology (e.g., VO_2max, ventilatory thresholds, etc.) do not provide sufficient information about the dynamic interactions among physiological systems and their common role in an integrated network. In this line, previously published works have shown a lower sensitivity of gold standards such as VO_2max compared with other coordinative variables able to determine dynamic interactions among physiological systems [104–107]. The different data analysis techniques described in the previous section have the potential to be used as novel evaluation tools to investigate interactions among physiological systems under exercise settings. More specifically, these techniques can lead to the development of new network-based biomarkers able to quantify how different key organ systems (e.g., brain, heart, skeletal muscles) coordinate and synchronize as a network during exercise and track how these network interactions change in response to fatigue and training. The use of new network-based biomarkers will break new ground in the study of multilevel inter-organ interactions and will provide new understanding of Basic Physiology and diverse exercise-related phenomena such as sports performance, fatigue, overtraining, or muscle-skeletal injuries.

4.2.1 Inter-muscular interactions

Inter-muscular coordination is defined as a distribution of muscle activation or force among individual muscles to produce a given combination of joint moments [108]. Therefore, neuromuscular control during exercise or activities of daily living is not limited to switching muscles on or off but includes fine-tuned control to select the appropriate muscle fiber types with precise timing and activation [109–111]. Techniques based on the frequency domain of the surface EMG [112, 113] are the most suitable to infer information on motor unit recruitment and muscle fiber since (i) the average conduction velocity of the active motor unit is related to fiber-type proportions, and (ii) the changes in the spectral properties are linked to the changes in the average conduction velocity. Inter-muscular coherence (IMC) is one of the most utilized methods to investigate inter-muscular interactions in the frequency domain—it estimates the amount of common neural input between two muscles during voluntary motor tasks [114]. Despite its clinical relevance to evaluate inter-muscular coordination, IMC has been recently questioned for its lack of potential to identify nonlinear dynamic coupling across frequencies [115] and, thus, ignore the interactions between distinct types of muscle fibers across muscles. Therefore, new data analysis approaches are needed to investigate the physiological mechanisms underlying cross-frequency network communication among distinct muscle fiber types across muscles during exercise.

4.2.2 Cortico-muscular interactions

Skeletal muscle activity is continuously modulated across physiologic states to provide coordination, flexibility, and responsiveness to body tasks and external inputs. Despite the central role the muscular system plays in facilitating vital body functions, the network of brain-muscle interactions required to control hundreds of muscles and synchronize their activation in relation to distinct physiologic states has not been sufficiently investigated. In this line, to identify and quantify the cortico-muscular interaction network and uncover basic features of neuro-autonomic control of muscle function, a recently published work [116] has investigated the coupling between synchronous bursts in cortical rhythms and peripheral muscle activation during sleep and wake. The findings demonstrate previously unrecognized basic principles of brain-muscle network communication and control and provide new perspectives on the regulatory mechanisms of brain dynamics and locomotor activation, with potential clinical implications for neurodegenerative, movement, and sleep disorders and for developing efficient treatment strategies. Further research is warranted to investigate cortico-muscular interactions during exercise and their changes in response to fatigue and different training methodologies.

4.2.3 Cardiorespiratory interactions

Previous research has demonstrated that the cardiac and respiratory systems exhibit three distinct forms of coupling: respiratory sinus arrhythmia (RSA), cardiorespiratory phase synchronization (CRPS), and time-delay stability (TDS) [76, 85, 101]. While RSA is a measure of amplitude modulation of the heart rate during the breathing cycle, CRPS and TDS characterize the temporal coordination between the cardiac and respiratory systems. Specifically, the CRPS reflects the degree of clustering of heartbeats at specific relative phases within each breathing cycle (despite continuous fluctuations in heart rate and in breathing intervals), and the TDS quantifies the stability of the time delay with which bursts in the activity in one system are consistently followed by corresponding bursts in the other system. The findings indicate that these three distinct and independent forms of cardiorespiratory coupling are of transient nature, with nonlinear temporal organization of intermittent "on" and "off" periods, even during the same episode of any given physiologic state (sleep stage), and that these coupling forms can simultaneously coexist.

In the context of exercise, cardiorespiratory coordination has been investigated through a Principal Components Analysis (PCA) performed on time series of cardiovascular and respiratory variables registered during cardiorespiratory exercise testing (expired fraction of O_2, expired fraction of CO_2, ventilation, systolic blood pressure, diastolic blood pressure, and heart rate). Cardiorespiratory coordination has been utilized to assess changes produced by different training programs [103, 105], testing manipulations [104, 106, 117, 118], nutritional interventions [107], and pathological conditions [119]. The main findings of this set of studies point toward a higher sensitivity and responsiveness of cardiorespiratory coordination to exercise effects compared with isolated cardiorespiratory parameters, such as VO_2max and other gold standard markers of aerobic fitness.

It should be noted that the aforementioned network-based biomarkers (intermuscular interactions, cortico-muscular interactions and cardiorespiratory interactions) can provide relevant information about how different key organ systems coordinate and synchronize as a network during exercise and track how these network interactions reorganize with accumulation of fatigue and in response to different training programs. However, these network-based biomarkers can only provide information at the organic (macroscopic) level by using equipment capable of recording continuous high-frequency physiological signals (time series of EEG, ECG, EMG). Therefore, the development of adequate technology able to register continuous and synchronous data extracted from different levels is needed to investigate the dynamics of physiological network interactions (i) not only at a macroscopic level, but also at lower levels of integration (i.e., cellular and subcellular); and (ii) among multilevel systems components—that is, capturing the synergies, embeddednes, and circular causality (bottom-up, top-down) between lower and upper levels.

5. Conclusions

Despite the fundamental discoveries, vast progress and achievements in the field of EP for over a century, the reductionist framework that has traditionally dominated research in the field has imposed limitations to the exploration and understanding of the regulatory mechanisms underlying complex exercise-related phenomena.

EP research, characterized by an inductive analytic mode of inquiry, has progressively evolved toward Biochemistry, Molecular Biology, Genetics, and OMICS technologies. Although such biology branches can be subjected to dynamical approaches, Molecular Exercise Physiology and Integrative Physiology keep focused on qüestionable non-dynamic bottom-up group-pooled statistical inferences.

Inspired by the field of Network Physiology and Complex Systems Science, Network Physiology of Exercise emerges to transform the theoretical assumptions, the research program and the practical applications of EP. The cybernetic Control Theory is replaced by Dynamic Systems Theory (DST), the centralized control of the CNS by a multilevel self-organization of body functions, and the static regulatory mechanisms by dynamic mechanisms with synergetic properties. The inductive analytical research, generalizing from group inter-individual inferences to intra-individual phenomena, is replaced by an inductive/deductive research based on intra-individual time series analysis techniques. Furthermore, it fills the gap of current research in Molecular Exercise Physiology, which is almost exclusively based on establishing bottom-up static statistical inferences from molecular data to the physiology of the entire person.

Network Physiology of Exercise focuses the research efforts on investigating the nested dynamics of the vertical (among levels) and horizontal physiological network interactions. The embeddedness of lower network levels in upper levels, the circular causality (bottom-up, top-down) among levels acting at different timescales and the emergence of nonlinear network phenomena are some of its genuine expected contributions. Network Physiology provides a wide range of data analysis techniques that have the potential to be utilized as novel evaluation tools to investigate interactions among physiological systems under exercise settings. These techniques can lead to the development of new network-based biomarkers (e.g., cardiorespiratory interactions, inter-muscular interactions, and cortico-muscular interactions) able to identify how different key organ systems coordinate and synchronize as a network during exercise and track how these network interactions change in response to different physiological states and exercise interventions. The use of new network-based biomarkers will open new and exciting horizons on exercise testing, will enrich Basic Physiology and diverse fields such as Exercise Physiology, Sports Medicine, Sports Rehabilitation, Sport Science, or Training Science, and will improve the understanding of diverse exercise-related phenomena such as sports performance, fatigue, overtraining, or sport injuries.

Acknowledgements

It includes funding information.

Author details

Natàlia Balagué[1*], Sergi Garcia-Retortillo[1,2,3], Robert Hristovski[4]
and Plamen Ch. Ivanov[3,5,6]

1 Complex Systems in Sport, INEFC Universitat de Barcelona (UB), Barcelona,
Spain

2 University School of Health and Sport (EUSES), University of Girona, Girona,
Spain

3 Keck Laboratory for Network Physiology, Department of Physics, Boston
University, Boston, MA, United States

4 Faculty of Physical Education, Sport and Health, Ss. Cyril and Methodius
University, Skopje, North Macedonia

5 Harvard Medical School and Division of Sleep Medicine, Brigham and Women's
Hospital, Boston, MA, United States

6 Institute of Solid State Physics, Bulgarian Academy of Sciences, Sofia, Bulgaria

*Address all correspondence to: nataliabalague@gmail.com

IntechOpen

References

[1] Balagué N, Hristovski R, Almarcha MC, García-Retortillo S, Ivanov P. Network physiology of exercise: Vision and perspectives. Frontiers in Physiology. 2020;**11**:611550. DOI: 10.3389/fphys.2020.611550

[2] Tipton CM. History of Exercise Physiology. Champaign, IL: Human Kinetics; 2014

[3] Hill AV, Long CHN, Lupton H. Muscular exercise, lactic acid and the supply and utilisation of oxygen. Proceedings of the Royal Society B. 1924;**96**:438-444

[4] Meyerhof O. The Chemical Mechanisms in the Skeletal Muscle and the Context With Performance Capacity and the Emergence of Heat [in German]. Berlin: Springer; 1930

[5] Kenney WL, Wilmore JH, Costill DL. Physiology of Sport and Exercise. 7th ed. Champaign. IL: Human Kinetics; 2019

[6] McArdle WD, Katch FI, Katch VL. Exercise Physiology: Nutrition, Energy, and Human Performance. Lippincott Williams & Wilkins; 2010

[7] Hristovski R, Aceski A, Balagué N, Seifert L, Tufekcievski A, Aguirre C. Structure and dynamics of European sports science textual contents: Analysis of ECSS abstracts (1996-2014). European Journal of Sport Science. 2017;**17**(1):19-29. DOI: 10.1080/17461391.2016.1207709

[8] Gomes C, Almeida JA, Franco OL, Petriz B. Omics and the molecular exercise physiology. Advances in Clinical Chemistry. 2020;**96**:55-84. DOI: 10.1016/bs.acc.2019.11.003

[9] Head GA. Integrative physiology: Update to the grand challenge. Frontiers in Physiology. 2020;**11**:489. DOI: 10.3389/fphys.2020.00489

[10] Sieck GC. Physiology in perspective: The importance of integrative physiology. Physiology. 2017;**32**:180-181. DOI: 10.1152/physiol.00009.2017

[11] Wackerhage H, editor. Molecular Exercise Physiology: An Introduction. London: Routledge; 2014

[12] Kitano H. Computational systems biology. Nature. 2002;**420**:206-210. DOI: 10.1038/nature01254

[13] Murtaza I, Majeed OL, Ubaid-ullah S. DNA Microrray: A miniaturized high throughput technology. International Journal of Engineering Research and Development. 2016;**12**:2278-2800

[14] Kohl M, Megger DA, Trippler M, Meckel H, Ahrens M, Bracht T, et al. A practical data processing workflow for multi-OMICS projects. Biochimica et Biophysica Acta. 2014;**1844**(1 Pt A):52-62. DOI: 10.1016/j.bbapap.2013.02.029

[15] Burniston JG, Chen YW, editors. Omics Approaches to Understanding Muscle Biology. New York: Springer; 2019

[16] Gonçalves LC, Bessa A, Freitas-Dias R, Luzes R, Werneck-de-Castro JPS, Bassini A, et al. A sportomics strategy to analyze the ability of arginine to modulate both ammonia and lymphocyte levels in blood after high-intensity exercise. Journal of the International Society of Sports Nutrition. 2012;**9**(1):30. DOI: 10.1186/1550-2783-9-30

[17] Hoffman NJ. Omics and exercise: Global approaches for mapping exercise biological networks. Cold Spring Harbor Perspectives in Medicine. 2017;7(10):a029884. DOI: 10.1101/cshperspect.a029884

[18] Hargreaves M. Fatigue mechanisms determining exercise performance:

Integrative physiology is systems biology. Journal of Applied Physiology. 2008;**104**:1541-1542

[19] Sturmberg JP. Embracing Complexity in Health. Cham: Springer; 2019

[20] Dasanayaka NN, Sirisena ND, Samaranayake N. The effects of meditation on length of telomeres in healthy individuals: A systematic review. Systematic Reviews. 2021;**10**(1):151. DOI: 10.1186/s13643-021-01699-1

[21] Tolahunase M, Sagar R, Dada R. Impact of yoga and meditation on cellular aging in apparently healthy individuals: A prospective, open-label single-arm exploratory study. Oxidative Medicine and Cellular Longevity. 2017;**7928981**:1-9. DOI: 10.1155/2017/7928981

[22] Lin J, Epel E. Stress and telomere shortening: Insights from cellular mechanisms. Ageing Research Reviews. 2022;**73**:101507. DOI: 10.1016/j.arr.2021.101507

[23] Balagué N, Torrents C, Hristovski R, Kelso JAS. Sport science integration: An evolutionary synthesis. European Journal of Sport Science. 2017;**17**(1):51-62

[24] Sturmberg JP, Martin CM. Handbook of Systems and Complexity in Health. New York: Springer. 2013. pp. 1-17

[25] Balagué N, Pol R, Torrents C, Ric A, Hristovski R. On the Relatedness and Nestedness of Constraints. Sports Medicine - Open. 2019;**5**(1). DOI: 10.1186/s40798-019-0178-z

[26] Greenhaff PL, Hargreaves M. 'Systems biology' in human exercise physiology: Is it something different from integrative physiology? The Journal of Physiology. 2011;**589**(Pt 5):

1031-1036. DOI: 10.1113/jphysiol.2010.201525

[27] Noble D. The music of life. Oxford: Oxford University Press; 2006

[28] Zierath JR, Wallberg-Henriksson H. Looking ahead perspective: Where will the future of exercise biology take us? Cell Metabolism. 2015;**22**:25-30

[29] World Health Organization. Global action plan on physical activity 2018-2030: More active people for a healthier world. Geneva, Switzerland: World Health Organization; 2019

[30] Enoka RM, Duchateau J. Translating fatigue to human performance. Medicine and Science in Sports and Exercise. 2016;**48**(11):2228-2238. DOI: 10.1249/MSS.0000000000000929. PMID: 27015386; PMCID: PMC5035715

[31] Noakes TD, Gibson SC, A. & Lambert, EV. From catastrophe to complexity: A novel model of integrative central neural regulation of effort and fatigue during exercise in humans. British Journal of Sports Medicine. 2004;**38**:511-514

[32] Amann M et al. Commentaries on viewpoint: Fatigue mechanisms determining exercise performance: Integrative physiology is systems physiology. Journal of Applied Physiology. 2008;**104**(1543-1546):2008. DOI: 10.1152/japplphysiol.90427.2008

[33] Marcora SM, Staiano W. The limit to exercise tolerance in humans: mind over muscle? European Journal of Applied Physiology. 2010;**109**:763-770. DOI: 10.1007/s00421-010-1418-6

[34] Allen DG, Westerblad H. Role of phosphate and calcium stores in muscle fatigue. The Journal of Physiology. 2001;**536**(Pt 3):657-665

[35] Amman M et al. Commentaries on viewpoint: Fatigue mechanisms

determining exercise performance: Integrative physiology is systems physiology. Journal of Applied Physiology. 2008;**104**(1543-1546):2008. DOI: 10.1152/japplphysiol.90427.2008

[36] Boyas S, Guével A. Neuromuscular fatigue in healthy muscle: Underlying factors and adaptation mechanisms. Annals of Physical and Rehabilitation Medicine. 2011;**54**(2):88-108

[37] Place N, Yamada T, Bruton JD, Westerblad H. Muscle fatigue: From observations in humans to underlying mechanisms studied in intact single muscle fibres. European Journal of Applied Physiology. 2010;**110**(1):1-15

[38] Brooks GA. Lactate as a fulcrum of metabolism. Redox Biology. 2020;**35**:101454. DOI: 10.1016/j.redox.2020.101454

[39] Ferguson BS, Rogatzki MJ, Goodwin ML, Kane DA, Rightmire Z, Gladden LB. Lactate metabolism: historical context, prior misinterpretations, and current understanding. European Journal of Applied Physiology. 2018;**118**(4):691-728. DOI: 10.1007/s00421-017-3795-6

[40] Balagué N, Hristovski R, Vainoras A, Vazquez P, Aragonés D. Psychobiological integration during exercise. In: Davids K, Hristovski R, Araújo D, Balagué N, Button CY, Passos P, editors. Complex Systems in Sport. London: Routledge; 2014. pp. 82-102

[41] Vázquez P, Hristovski R, Balagué N. The path to exhaustion: Time-variability properties of coordinative variables during continuous exercise. Frontiers in Physiology. 2016;7:37. DOI: 10.3389/fphys.2016.00037

[42] Vázquez P, Petelczyc M, Hristovski R, Balagué N. Interlimb coordination: A new order parameter and a marker of fatigue during

quasi-isometric exercise? Frontiers in Physiology. 2021;**11**:612709. DOI: 10.3389/fphys.2020.612709 I.F: 3.367

[43] Bechtel W, Abrahamsen AA. Thinking dynamically about biological mechanisms: Networks of coupled oscillators. Foundations of Science. 2013;**18**:707-723. DOI: 10.1007/s10699-012-9301-z

[44] Wilson EO. Consilience: The Unity of Knowledge. New York: Knopf; 1998

[45] Neufer PD, Bamman MM, Muoio DM, Bouchard C, Cooper DM, Goodpaster BH, et al. Understanding the cellular and molecular mechanisms of physical activity-induced health benefits. Cell Metabolism. 2015;**22**(1): 4-11. DOI: 10.1016/j.cmet.2015.05.011 Epub 2015 Jun 11. PMID: 26073496

[46] Hopkins WG. Effects went away at the 2013 annual meeting of the European College of Sport Science. Sportscience. 2013;**17**:1-12

[47] Hristovski R, Venskaitytė E, Vainoras A, Balagué N, Vazquez P. Constraints- controlled metastable dynamics of exercise-induced psychobiological adaptation. Medicina (B. Aires). 2010;**46**:447-453. DOI: 10.3390/medicina46070064

[48] Armstrong L, Bergeron M, Lee E, Mershon J, Armstrong E. Overtraining syndrome as a complex system. Frontiers in Network Physiology. 2021 (accepted)

[49] Pol R, Hristovski R, Medina D, Balagué N. From micro- to macroscopic injuries: Applying the complex systems dynamic approach to sports medicine. British Journal of Sports Medicine. 2018;**0**:1-8. DOI: 10.1136/bjsports-2016-097395

[50] Bassingthwaighte JB, Liebovitch LS, West BJ, Stanley HE. Fractal physiology. Physics Today. 1995;**48**:66-66. DOI: 10.1063/1.2808299

[51] Lambert EV. Complex systems model of fatigue: integrative homoeostatic control of peripheral physiological systems during exercise in humans. British Journal of Sports Medicine. 2005;**39**:52-62. DOI: 0.1136/bjsm.2003.011247

[52] Kelso JAS. Dynamic Patterns. Cambridge: MIT Press; 1997

[53] Molenaar PCM. A manifesto on psychology as idiographic science: bringing the person Back into scientific psychology, this time forever. Measurement: Interdisciplinary Research and Perspectives. 2004;**2**:201-218. DOI: 10.1207/s15366359mea0204_1

[54] Rose LT, Rouhani P, Fischer KW. The science of the individual. Mind, Brain, and Education. 2013;**7**:152-158. DOI: 10.1111/mbe.12021

[55] Rose LT. The End of Average: How to Succeed in a World that Values Sameness. USA: Harper Collins; 2016

[56] Topa H, Honkela A. GP rank: And R package for detecting dynamic elements from genome-wide time series. BMC Bioinformatics. 2018;**19**:367. DOI: 10.1186/s12859-018- 2370-4

[57] Elbich DB, Molenaar PC, Scherf KS. Evaluating the organizational structure and specificity of network topology within the face processing system. Human Brain Mapping. 2019;**40**(9):2581-2595

[58] Gates KM, Lane ST, Varangis E, Giovanello K, Guiskewicz K. Unsupervised classification during time-series model building. Multivariate Behavioral Research. 2017;**52**(2):129-148

[59] Beltz AM, Gates KM. Network mapping with GIMME. Multivariate Behavioral Research. 2017;**52**(6):789-804

[60] Henry TR, Feczko E, Cordova M, Earl E, Williams S, Nigg JT, et al. Comparing directed functional connectivity between groups with confirmatory subgrouping GIMME. NeuroImage. 2019;**188**:642-653

[61] Noble R, Tasaki K, Noble PJ, Noble D. Biological relativity requires circular causality but not symmetry of causation: So, where, what and when are the boundaries? Frontiers in Physiology. 2019;**10**:827. DOI: 10.3389/fphys.2019.00827

[62] Tarasov VE. Self-organization with memory. Communications in Nonlinear Science and Numerical Simulation. 2019;**72**:240-271. DOI: 10.1016/j.cnsns.2018.12.018

[63] Kelso JAS. Synergies: Atoms of brain and behavior. In: Sternad D, editor. Progress in Motor Control. Advances in Experimental Medicine and Biology. Vol. 629. Boston, MA: Springer; 2009. DOI: 10.1007/978-0-387-77064-2_5

[64] Kelso JAS. Unifying large- and small-scale theories of coordination. Entropy. 2021;**23**:537. DOI: 10.3390/e23050537

[65] Edelman GM, Gally JA. Degeneracy and complexity in biological systems. Proceedings of the National Academy of Sciences of the United States of America. 2001;**98**:13763-13768. DOI: 10.1073/pnas.231499798

[66] Latash ML. Human movements: synergies, stability, and agility. In: Siciliano B, Khatib O, editors. Springer Tracts in Advanced Robotics. Berlin: Springer Verlag); 2019. pp. 135-154

[67] Pedersen BK, Febbraio MA. Muscle as an endocrine organ: Focus on muscle-derived interleukin- 6. Physiological Reviews. 2008;**88**:1379-1406

[68] Sallam N, Laher I. Exercise Modulates Oxidative Stress and Inflammation in Aging and Cardiovascular Diseases. Oxidative

Medicine and Cellular Longevity. 2016;**2016**:7239639. DOI: 10.1155/2016/7239639

[69] Bovier A, Den Hollander F. Metastability: A Potential-Theoretic Approach. New York: Springer; 2016

[70] Kiebel SJ, Daunizeau J, Friston KJ. A hierarchy of time-scales and the brain. PLoS Computational Biology. 2008;**4**(11):e1000209. DOI: 10.1371/journal.pcbi.1000209

[71] Pollak GH, Chin WC. Phase Transitions in Cell Biology. London: Springer; 2008

[72] Barabási A-L, Oltvai ZN. Network biology: Understanding the cell's functional organization. Nature Reviews Genetics. 2004;**5**(2):101-113. DOI: 10.1038/nrg1272

[73] Balleza E, Alvarez-Buylla ER, Chaos A, Kauffman S, Shmulevich I, Aldana M. Critical dynamics in genetic regulatory networks: Examples from four kingdoms. PLoS One. 2008;**3**:e2456. DOI: 10.1371/journal.pone.0002456

[74] Micheel CM, Nass SJ, Omenn GS, editors. Evolution of Translational Omics: Lessons Learned and the Path Forward. Washington (DC): National Academies Press (US); 2012

[75] Flavahan et al. Science. 2017;**357**:eaal2380

[76] Bartsch RP, Ivanov PC. Coexisting forms of coupling and phase-transitions in physiological networks. In: Mladenov VM, Ivanov PC, editors. Communications in Computer and Information Science. Vol. 438. Cham: Springer; 2014. pp. 270-287

[77] Ivanov P.C., Wang J.W.J.L., Zhang X., Chen B. (2021b) The new frontier of Network Physiology: Emerging physiologic states in health and disease from integrated organ network interactions. In: Wood D.R., de Gier J., Praeger C.E., Tao T. (eds) 2019-20 MATRIX Annals. MATRIX Book Series, vol 4. Springer, Cham. 10.1007/978-3-030-62497-2_12

[78] Ivanov PC, Bartsch RP. Physiologic systems dynamics, coupling and network interactions across the sleep-wake cycle. In: Murillo-Rodríguez E, editor. Methodological Appraches for Sleep and Vigilance Research. Cambridge: Academic Press; 2022. pp. 59-100. DOI: 10.1016/C2020-0-01745-7

[79] Ivanov PC. The new field of network physiology: Building the human Physiolome. Frontiers in Network Physiology. 2021b;**1**:711778. DOI: 10.3389/fnetp.2021.711778

[80] Bashan A, Bartsch RP, Kantelhardt JW, Havlin S, Ivanov PC. Network physiology reveals relations between network topology and physiological function. Nature Communications. 2012;**3**:702. DOI: 10.1038/ncomms1705

[81] Bartsch RP, Liu KKL, Bashan A, Ivanov PC. Network physiology: how organ systems dynamically interact. PLoS One. 2015;**10**:e0142143. DOI: 10.1371/journal.pone.0142143

[82] Liu KK, Bartsch RP, Ma QD, Ivanov PC. Major component analysis of dynamic networks of physiologic organ interactions. Journal of Physics Conference Series. 2015b;**640**:012013. DOI: 10.1088/1742-6596/640/1/012013

[83] Ivanov PC, Liu KKL, Bartsch RP. Focus on the emerging new fields of network physiology and network medicine. New Journal of Physics. 2016;**18**:100201. DOI: 10.1088/1367-2630/18/10/100201

[84] Ivanov PC, Liu KK, Lin A, Bartsch RP. Network physiology: from neural plasticity to organ network interactions. In: Mantica G, Stoop R,

Stramaglia S, editors. Emergent complexity from nonlinearity, in physics, engineering and the life sciences. Cham: Springer; 2017. pp. 145-165

[85] Bartsch RP, Schumann AY, Kantelhardt JW, Penzel T, Ivanov PC. Phase transitions in physiologic coupling. Proceedings of the National Academy of Sciences of the United States of America. 2012;**109**:10181-10186. DOI: 10.1073/pnas.1204568109

[86] Chen Z, Hu K, Stanley HE, Novak V, Ivanov, PC. Cross-correlation of instantaneous phase increments in pressure-flow fluctuations: Applications to cerebral autoregulation. Physical Review E. 2006;**73**:031915. DOI: 10.1103/PhysRevE.73.031915

[87] Ivanov PC, Ma QD, Bartsch RP, Hausdorff JM, Amaral LAN, Schulte-Frohlinde V, et al. Levels of complexity in scale-invariant neural signals. Physical Review E. 2009;**79**:041920. DOI: 10.1103/PhysRevE.79.041920

[88] Bechtel W, Abrahamsen A. Dynamic mechanistic explanation: computational modeling of circadian rhythms as an exemplar for cognitive science. Studies in History and Philosophy of Science. 2010;**41**:321-333. DOI: 10.1016/j.shpsa.2010.07.003

[89] Meyer R. The non-mechanistic option: defending dynamical explanations. The British Journal for the Philosophy of Science. 2020;**71**:959-985. DOI: 10.1093/bjps/axy034

[90] Suchecki K, Eguíluz V, San Miguel M. Voter model dynamics in complex networks: Role of dimensionality, disorder, and degree distribution. Physical Review E. 2005;**72**(3):036132. DOI: 10.1103/PhysRevE.72.036132

[91] Latané B. Dynamic social impact: The creation of culture by communication. Journal of Communication. 1996;**46**(4):13-25. DOI: 10.1111/j.1460-2466.1996. tb01501.x

[92] Vallacher RR, Nowak A. Dynamical social psychology: On the complexity and coordination in interpersonal relations. In: Uhl-Bien M, Marion R, Hanges P, editors. Toward a complexity theory of leadership and organizing. Charlotte, NC: Information Age; 2007. pp. 49-81

[93] Meeusen R, Duclos M, Foster C, Fry A, Gleeson M, Nieman D, et al. Prevention, diagnosis, and treatment of the overtraining syndrome: Joint consensus statement of the European College of Sport Science and the American College of Sports Medicine. Medicine and Science in Sports and Exercise. 2013;**45**:186-205. DOI: 10.1249/MSS.0b013e318279a10a

[94] Gershenson C, Fernández N. Complexity and information: Measuring emergence, self-organization, and homeostasis at multiple scales. Complexity. 2012;**18**:29-44. DOI: 10.1002/cplx.21424

[95] Van Orden GC, Holden JC, Turvey MT. Self-organization of cognitive performance. Journal of Experimental Psychology. General. 2003;**132**:331-350. DOI: 10.1037/0096-3445.132.3.331

[96] Delignieres D, Marmelat V. Fractal fluctuations and complexity: current debates and future challenges. Critical Reviews in Biomedical Engineering. 2012;**40**:485-500. DOI: 10.1615/CritRevBiomedEng.2013006727

[97] Almurad ZMH, Roume C, Blain H, Delignières D. Complexity matching: Restoring the complexity of locomotion in older people through arm-in-arm walking. Frontiers in Physiology. 2018;**9**:1766. DOI: 10.3389/fphys.2018.01766

[98] Venhorst A, Micklewright D, Noakes TD. Towards a three-dimensional framework of centrally regulated and goal-directed exercise behaviour: a narrative review. British Journal of Sports Medicine. 2018;**52**:957-966. DOI: 10.1136/bjsports-2016-096907

[99] Pol R, Balagué N, Ric A, Torrents C, Kiely J, Hristovski R. Training or synergizing? Complex systems principles change the understanding of sport processes. Sports Medicine. 2020;**6**:28. DOI: 10.1186/s40798-020-00256-9

[100] Ivanov PCh, Wang JWJL, and Zhang X. Signal processing in Network Physiology: quantifying network dynamics of organ interactions. 2020 28th European Signal Processing Conference (EUSIPCO), Amsterdam, Netherlands, 2021a, pp. 945-949

[101] Bartsch RP, Liu KKL, Ma QD, Ivanov PC. Three independent forms of cardio- respiratory coupling: Transitions across sleep stages. Computers in Cardiology. 2014;**41**:781-784

[102] Hacken H. Information and self-organization. In: A Macroscopic Approach to Complex Systems. 3rd ed. New York, NY: Springer; 2010. pp. 69-152

[103] Balagué N, González J, Javierre C, Niño O, Alamo J, Aragonés D, et al. Cardiorespiratory coordination after training and detraining. Principal component analysis approach. Frontiers in Physiology. 2016;**7**:35. DOI: 10.3389/fphys.2016.00035

[104] Garcia S, Javierre C, Hristovski R, Ventura JL, Balagué N. Cardiorespiratory coordination in repeated maximal exercises. Frontiers in Physiology. 2017;**8**:387. DOI: 10.3389/fphys.2017.00387

[105] Garcia-Retortillo S, Gacto M, O'Leary TJ, et al. Cardiorespiratory coordination reveals training-specific physiological adaptations. European Journal of Applied Physiology. 2019a;**119**:1701-1709. DOI: 10.1007/s00421-019-04160-3

[106] Garcia-Retortillo S et al. Physiological Measurement. 2019b;**40**:084002

[107] Esquius L, Garcia-Retortillo S, Balagué N, Hristovski R, Javierre C. Physiological- and performance-related effects of acute olive oil supplementation at moderate exercise intensity. Journal of the International Society of Sports Nutrition. 2019;**16**:12. DOI: 10.1186/s12970-019-0279-6

[108] Prilutsky BI. Coordination of two- and one-joint muscles: Functional consequences and implications for motor control. Motor Control. 2000;**4**(1):1-44. DOI: 10.1123/mcj.4.1.1

[109] Kristiansen M, Samani A, Madeleine P, Hansen EA. Effects of 5 weeks of bench press training on muscle synergies: A randomized controlled study. Journal of Strength and Conditioning Research. 2016;**30**(7): 1948-1959

[110] Lin A, Liu KKL, Bartsch RP, Ivanov PC. Dynamic network interactions among distinct brain rhythms as a hallmark of physiologic state and function. Communications Biology. 2020;**3**:197

[111] Lombardi F, Wang JWJL, Zhang X, Ivanov PC. Power-law correlations and coupling of active and quiet states underlie a class of complex systems with self-organization at criticality. EPJ Web of Conferences. 2020;**230**:00005

[112] Garcia-Retortillo RR, Ivanov PC. Spectral dynamics of muscle fiber activation in response to exercise and acute fatigue. IEEE EMBS International Conference on Biomedical and Health Informatics (BHI). 2021;**2021**:1-4

[113] Garcia-Retortillo S, Rizzo R, Wang JWJL, Sitges C, Ivanov PC. Universal spectral profile and dynamic evolution of muscle activation: a hallmark of muscle type and physiological state. Journal of Applied Physiology. 2020;**129**(3):419-441

[114] Ushiyama J, Ushiba J. Resonance between cortex and muscle: A determinant of motor precision? Clinical Neurophysiology. 2013;**124**:5-7. DOI: 10.1016/j. clinph.2012.08.004

[115] Yang Y, Dewald JP, van der Helm FC, Schouten AC. Unveiling neural coupling within the sensorimotor system: Directionality and nonlinearity. European Journal of Neuroscience. 2018;**48**(7):2407-2415

[116] Rizzo R, Zhang X, Wang JWJL, Lombardi F, Ivanov PC. Network physiology of Cortico– Muscular interactions. Frontiers in Physiology. 2020;**11**:558070

[117] Zebrowska M, Garcia-Retortillo S, Sikorski K, Balagué N, Hristovski R, Javierre C, et al. Decreased coupling among respiratory variables with effort accumulation. Europhysics Letters. 2020;**132**:28001 www.epljournal.org. DOI: 10.1209/0295-5075/132/28001

[118] Uryumtsev DY, Gultyaeva VV, Zinchenko MI, Baranov VI, Melnikov VN, Balioz NV, et al. Effect of acute hypoxia on cardiorespiratory coherence in male runners. Frontiers in Physiology. 2020;**11**:630. DOI: 10.3389/fphys.2020.00630

[119] Oviedo GR, Garcia-Retortillo S, Carbó-Carreté M, Guerra-Balic M, Balagué N, Javierre C, et al. Cardiorespiratory coordination during exercise in adults with down syndrome. Frontiers in Physiology. 2021;**12**:704062. DOI: 10.3389/fphys.2021.704062

Chapter 9

Energy Cost of Walking and Running

Vaclav Bunc

Abstract

Walking and running are the basic means of influencing an individual's condition, his or her health and fitness. Due to the fact that various forms of physical load are used in movement training, the cause must be described by a single number, which reflects the volume, intensity, and form of physical load. One of the possibilities is to determine the energy cost (EC) of the applied physical activities. Possibilities of evaluation of EC in laboratory and field conditions using the speed of movement allow to streamline movement training. To achieve the desired lasting effect, it is necessary that the total EC exceeds the so-called stimulus threshold, that is, the subject of physical training must reach a certain minimum level of total EC of applied physical training. The total energy content of exercise allows you to design individual exercise programs. In the paper, we present the relationships between energy and speed of movement for the most commonly used physical activities to increase fitness in people without regular physical training–walking and running in different age groups and for men and women and the principles of design of movement interventions using this parameter, as well as the implemented programs and their effect.

Keywords: walking, running, energy, intensity of exercise, energy cost of movement, movement economy

1. Introduction

Determining the energy intensity of physical activity is a basic problem in evaluating the impact of this activity on the human body, either in terms of certain civilization diseases prevention, or in terms of increasing the functional (physical) fitness of people, or to assess the body's response to a given type of load.

The goal of all non-pharmacological intervention programs – movement programs in primary and secondary prevention is to determine the minimum amount of exercise load that will cause the necessary persistent changes in the state of the organism [1–3].

In general, the amount of energy that an organism consumes in a given physical activity is directly proportional to the intensity of that activity [1, 4, 5]. Throughout the range of movement load intensities–movement speed v, the energy E required to provide locomotor activity is proportional to the power of the velocity of movement. This relationship is generally nonlinear over the entire range of load intensities v and can be described by the following Equation [2, 6, 7].

$$E = c * v^n \qquad (1)$$

Where c is the energy cost of movement, n for human movement ranges from 1 to 3 and expresses the density of the environment in which the movement is realized. The higher the density of the environment, the higher the value of n. The coefficient c characterizes the economics of motion and it is true that the lower its value, the better the economics of movement and the better its technique [2, 8].

The energy demands for submaximal movement intensity (i.e. movement economy) can be quantified by calculating the steady-state VO_2, expressed with respect to body mass and time, for a standardized, submaximal movement intensity [1, 9]. Because this variable represents an aerobic need for physical activity, ATP resynthesis from ADP must be paid exclusively from substrates stored in the body and oxygen obtained from pulmonary ventilation and not from substantial protein catabolism. In untrained individuals, research has shown that at low to moderate speeds, steady state oxygen consumption is reached in approximately 3 minutes. Trained individuals reach a steady state earlier than untrained individuals. Although the existence of steady state is limited by a number of methodological limitations, this steady state can also be demonstrated by a non-increasing accumulation of lactate in the blood and RER lower than 1.00. All this is significantly influenced by diet, where in the case of predominant protein intake, the RER is less than one [1].

With a constant speed or running, at submaximal exercise intensities, the relationship between E and speed of running v is linear Energy necessary to proceed at a given running speed can be regarded as the product of c coefficient times the speed itself

$$E = c * v \qquad (2)$$

Where c is in $J.kg^{-1}.m^{-1}$ and running speed in $m.s^{-1}$, thus yielding energy E in $W.kg^{-1}$. the range of linearity depends not only on the actual training state, but also on the metabolic state, age, sex, and speed potential of the subjects studied [6, 10].

The linearity of this dependence depends on the subject's training state. For running, it is in the range of 20–80% of the maximum intensity of movement in the untrained, and in the range of 10–90% of the maximum intensity in the trained [10].

Direct measurement of energy during real physical activity is relatively complicated. For practical reasons, we have often expressed E as oxygen uptake for the activity. In these cases, it has been convenient to express c in $J.kg^{-1}.m^{-1}$ and running speed in $m.min^{-1}$, to obtain E in more customary units $ml.kg^{-1}.m^{-1}$. Thus the last equation may be rearranged as follows:

$$VO_2 = c * v \qquad (3)$$

Under aerobic conditions, since the energy E can be identified with VO_{2max}, and in the submaximal range of intensities v, the last equation becomes:

$$f * VO_{2max} = c * v \qquad (4)$$

where f is the fraction of VO_{2max} which may be utilized over a prolonged period of time [2, 11]. The duration of the competition and thus also the performance in training is obviously a decisive factor in determining the magnitude off. It is larger the longer the duration of the competitive performance [2, 6, 12].

The movement speed thus may be calculated as follows

$$v = f * VO_{2max} * c^{-1} \qquad (5)$$

It follows from the above relationship that the better the economy of movement - the lower c, the higher the speed the individual can move.

2. Energy cost of movement

Coefficient of movement energy cost during running (expressed in $J.kg^{-1}.m^{-1}$), and indicates how much energy is needed to transfer the body mass of 1 kg to the distance of 1 m. It holds that the better the economy of movement, the lower the values of c we find. In our older study, we found the following values of the coefficient c for different sports, gender, and age. The value of this coefficient ranges from 3.5 for highly trained runners on middle distances to values of about 4.2 for untrained people. For example for men and women respectively in adult middle distance runners C = 3.57 +/− 0.15 and 3.65 +/− 0.20, in adult long-distance runners C = 3.63 +/− 0.18 and 3.70 +/− 0.21, in adult canoeists C = 3.82 +/− 0.34 and 3.80 +/− 0.24, in young middle-distance runners C = 3.84 +/− 0.18 and 3.78 +/− 0.26 and in young long-distance runners C = 3.85 +/− 0.12 and 3.80 +/− 0.24 [2]. This similarity may be explained by the similar training states of both sexes, resulting from the intense training which did not differ in its relative intensity and frequency between the groups of men and women. Bunc and Heller [2] found a negative relationship was found between the energy cost of running and maximal oxygen uptake (VO_{2max}) expressed relative to body mass (for men r = −0.471, p < 0.001; for women r = −0.589, p < 0.001). Thus, the better the adaptation to a given movement load, the higher the values of the maximum consumption an individual can achieve and the lower the density of the coefficient c, and the better the technique of movement.

The energy demands of higher stride frequency at a given speed are frequently cited as the most plausible explanation for the higher energy cost of movement, for the higher coefficient c. This concept is based on the assumption that the energy required to move body mass should directly reflect the muscle tension created by each stride [13].

If we assume that differences in stored energy are not significant between different subjects [1, 2] then we may conclude that in the non-trained subjects with increasing body fat percentage, and generally with increasing body mass, the coefficient of the energy cost of movement increases, and the energy cost of movement decreases when training state decreases. The prerequisites for using stored chemical energy for moving are reduced and thus the moving capacity and movement economy decline. For improving of predispositions for moving (transfer of body mass), it is necessary to reduce body mass to the subject's "optimal" body mass [4].

3. Energy cost of running

Movement economy c, which has traditionally been measured as the oxygen cost of running at a given velocity, has been accepted as the physiological criterion for 'efficient' performance and has been identified as a critical element of overall distance running performance [4, 6, 7, 11, 14]. It follows from the above that there is a relationship between the mechanics of running and its energy intensity, but previous research does not allow to determine a clear biomechanical profile of a runner with a high economy of movement - the high technique of movement. Through movement training, individuals seem to be able to adapt to achieve movement as economically as possible and minimize energy degradation [15]. Information in the literature suggests that biomechanical factors are likely to contribute to a better economy in any runners [16].

Atropometric parameters can significantly affect the biomechanics of movement, movement technique, and its energy intensity. These include height, ponderal index, and ectomorphic or ectomesomorphic figure; body fat percentage; foot morphology, pelvic size, foot size, and shape [13].

The economy of running can be influenced by movement patterns of running, their kinematics. These factors include the length of the step, which is freely chosen depending on the current fatigue; vertical oscillation of the center of gravity of the body; knee angle during the swing; range of motion, angular velocity of plantar flexion during toe-off; arm movement with smaller amplitude; peak ground reaction forces; rotation of the arms in the transverse plane; angular deflection of the hips and shoulders about the polar axis in the transverse plane; and efficient use of stored elastic energy [7, 17, 18]. Other factors that can significantly affect the running economy are: the shoes mainly their weight and the elasticity of the sole; higher share of higher and high training intensities of training history; and medium flexibility base. This information can be crucial in identifying talents for medium and long-distance running. At higher levels of training, it is likely that "natural selection" tends to eliminate athletes who have either failed to inherit or develop traits that support the economy of movement [19, 20].

It turns out that intra-individual variations in running economics range between 2% and 11% for a particular speed. Most of these variations are probably due to biological measurement error [21]. While the sources do not support gender differences in movement economics, data from some studies suggest that men may have better movement economics than women due to more muscle mass and less body fat. The economics of running change depending on age, depending on the amount of physical training completed. Pre-adolescent children have a worse economy than older children and adults, while older adults show the same trend compared to younger counterparts [9]. Air resistance at higher speeds fundamentally affects the economy of movement. Running on a treadmill at speeds higher than 13 km.h^{-1} due to air resistance significantly underestimates the cost of energy intensity compared to running speeds at the same speed in the field [9]. Oxygen consumption increases as a result of the "Q 10 effect" [22, 23]. There is also no consensus on the impact of different types and intensities of training on running economics, and significant differences in economics between long-distance runners who undergo the same load (eg track) suggest that non-training factors may also affect the running economy, such as the amount and type of muscle fibers [13, 20, 24].

From a study by Black et al. [19] show that anthropometric parameters and body composition are important predictors of running economics. Relative slenderness indices, especially segment perimeters, have been commonly associated with running economy, suggesting that a slimmer individual can be expected to expend less energy and thus be more economical at any given speed [25]. It should be noted that the amount of energy available for physical activity stored in the body per kg of free fat mass is practically the same for virtually all persons of the same sex. The importance of running economics in medium and long-distance running, we recommend to trainers, applied practitioners, and athletes to evaluate anthropometric parameters and body composition as part of the evaluation of training. This is especially important in identifying talented athletes and preparing top athletes to achieve maximum individual performance [26].

Studies comparing different groups of runners with different training and focus have shown that the maximum differences in energy intensity between runners are around 20%. Factors influencing the value of coefficient c include body dimensions: body height and weight, the architecture of the lower calves, mostly the length of the calcaneal tuberosity, which are responsible for 60–80% variability of this coefficient. Children have higher c values than adults. This can be explained by their higher resting metabolism, lower running technique, and lower leg/leg ratio [9, 27]. The storage of elastic energy and its reuse also contributes to the variability of c. The coefficient c increases with the increasing speed of movement due to the increase in mechanical work is blunted to speed of 6–7 m.s^{-1} by increasing the vertical stiffness and shortening the contact time with the ground. Fatigue caused by prolonged or intense running is associated with up to a 10% increase in c; the influence of metabolic and biomechanical factors on the energy intensity of running remains unclear. Women show c similar to men of similar body weight, despite differences in running technique. The higher performance of black African endurance athletes is probably related to their leg architecture and better elastic storage and reuse of elastic energy [20].

Speed and movement techniques are considered to be the main sources of changes in the energy intensity of running in individuals with different body masses. The linear dependence of energy on running speed is approximately up to a speed of 3.6 m.s^{-1}. In the case of higher speeds, this dependence is nonlinear. At6speeds higher than 3.6 m.s^{-1}, runners are less likely to achieve aerobic performance - steady state oxygen consumption [28].

Walking and running are the basic means of influencing an individual's condition, his health and fitness. Possibilities of evaluation of energy intensity in laboratory and field conditions using the speed of movement, allows to streamline movement training. Energy intensity allows you to design individual exercise programs, for example, for the needs of primary and secondary prevention of obesity, cardiovascular disease, reducing the impact of current lifestyles, etc. [3, 4, 29].

Human locomotion is characterized by two principal gaits, walking and running. This makes it possible to move either at a slow speed for long periods of time or at over 10 m.s^{-1} during a sprint [30]. The basic features of both locomotion modes are the same: each step represents one posture phase and one swing phase, but then they differ because the foot controls have two separate operating modes for walking and running. The timing of each phase of the movement is different. The frequency of steps is usually lower when walking than when running, so the contact time with the surface of each foot is longer when walking and shorter when running, while the swing shows the opposite trend. When walking, there is always at least one foot on the ground, while running there is a flight phase where both feet are above the ground and the amplitudes of the contractions of the flexor and extensor muscles during the two phases of the step are different [31, 32].

Studies examining the interaction between stride length, energy absorption, and impact attenuation have only been performed on level ground. Stepped running places unique demands on the musculoskeletal system compared to running on a plane, resulting in differences in physiological requirements and the kinematics and kinetics of the run [5]. Downhill running is associated with greater impact magnitudes and increased energy absorption when compared to level running [5]. The increased eccentric muscular work required to absorb more energy during downhill running may also be associated with muscle damage and delayed onset muscle soreness (DOMS), which negatively affects running performance. In contrast, uphill running is more energetically costly than level or downhill running [32] but is associated with lower impact magnitudes and reduced lower extremity energy absorption, especially when compared to downhill running [32]. Step length and frequency are also known to change during graded running [5], and step length manipulation may aid in understanding the injury and performance implications of these natural changes to preferred step length [18].

The evaluation of the energy intensity of running is a suitable criterion for examining the efficiency of mechanical work, evaluation of movement technique, and analysis of endurance performance during endurance running [24, 33].

Once energy cost values (VO_2 and caloric expenditure) are standardized using bodyweight, the primary determinant of energy cost was the speed of movement [1, 33]. The derived generalized models make it possible to determine both VO_2 ($ml.kg^{-1}.min^{-1}$) and the energy intensity ($kcal.kg^{-1}.min^{-1}$) of walking and running. The relationship between walking speed or number of steps and the energy intensity of walking is parabolic, while the relationship between running speed and energy intensity of running is in the range of about 20–80% for untrained and 20–90% for trained line runners is linear [34]. Neither age nor body height significantly improved the prediction of the energy cost of movement from its speed.

In practice, results of spiroergometric surveys often need to be checked using the relationship between oxygen consumption and movement speed. The relationship between energy and speed of running may be used as a linear form as follows

$$VO_2.kg^{-1} \left(ml.kg^{-1}. min^{-1}\right) = a * v \left(km.h^{-1}\right) + b \tag{6}$$

Where a and b are constants that depend generally on the training status, sex, age and speed, and strength predisposition.

Many equations can be found in the literature for predicting energy expenditure during walking or running. Not only can the amount of energy that was "burned" during a training unit be determined, but often these relationships are implemented in miniaturized electronic devices that provide the user with relevant data on the energy intensity of the physical activity performed. At the same time, it should be noted that the energy estimation error from walking or running speed is around 10% and these relationships can be used for so-called biological testing of spiroergometric analyzers. We include in the text those prediction equations which are currently the most frequently used and which provide relevant information for a given population. We have chosen the following tables and equations because they have often been cited in the literature. The ACSM Equation [35] was used because most exercise physiologists know the ACSM guidelines. McArdle's [33] walking and running tables have been used because they are found in commonly used exercise physiology textbooks and are often used by researchers in the field to estimate energy expenditure. Other equations were chosen because they were cited in the literature and provided additional estimates of walking and running. The prediction formulas that were used are listed below:

ACSM [34]:

$$Running.\dot{V}O_2 \left(mL{\cdot}kg^{-1}{\cdot} min^{-1}\right) = 0.2 \left(m{\cdot}s^{-1}\right) \\ + 0.9 \left(m{\cdot}s^{-1}\right) (fractional\ grade) + 3.5 \tag{7}$$

$$Walking.\dot{V}O_2 \left(mL{\cdot}kg^{-1}{\cdot} min^{-1}\right) = 0.1 \left(m{\cdot}s^{-1}\right) \\ + 1.8 \left(m{\cdot}s^{-1}\right) (fractional\ grade) + 3.5 \tag{8}$$

Bunc & Heller [14]
Running men

$$VO_2.kg^{-1} \left(mL{\cdot}kg^{-1}{\cdot} min^{-1}\right) = 3.749 * v \left(km.h^{-1}\right) - 2.133 \tag{9}$$

Bunc et al. [12]
Running women

$$VO_2.kg^{-1} \left(mL\cdot kg^{-1}\cdot min^{-1}\right) = 3.549 * v \left(km.h^{-1}\right) + 3.008 \tag{10}$$

Bunc & Dlouhá [34]
Walkimg

$$VO_2.kg^{-1} \left(mL\cdot kg^{-1}\cdot min^{-1}\right) = 3.207 * v \left(km.h^{-1}\right) - 1.777 \tag{11}$$

$$VO_2.kg^{-1} \left(mL\cdot kg^{-1}\cdot min^{-1}\right) = -0.108 * v \left(km.h^{-1}\right) \\ + 0.379 * v^2 \left(km.h^{-1}\right) + 4.503 \tag{12}$$

McArdle [33]:
McArdle's tables are available in the referenced text.
Van der Walt and Wyndham [36]:
Walking

$$\dot{V}O_2 \left(L\cdot min^{-1}\right) = 0.00599 * M + 0.000366 * M * V^2 \tag{13}$$

Running.

$$\dot{V}O_2 \left(l\cdot min^{-1}\right) = -0.419 + 0.03257 * M + 0.000117 * M * V^2 \tag{14}$$

Pandolf, Givoni & Goldman [37]:

$$W \left(J\cdot s^{-1}\right) = 1.5\,M + 2.0 * (M + L) * (L/M)^2 + n * (M + L) * \left[1.5 * v^2 + 0.35 * v * G\right] \tag{15}$$

M = body mass (kg), L = load carried, v = velocity ($m\cdot s^{-1}$), G = grade, and n is the terrain factor. For unloaded, level walking on a track or treadmill, the following formula is used:

$$W \left(1\,J\cdot s^{-1}\right) = 1.5 * W + 1.5 * v^2 * W \tag{16}$$

Léger & Mercier [38]:

$$\dot{V}O_2 \left(ml.kg^{-1}\cdot min^{-1}\right) = 2.209 + 3.1633 * v \left(km.h^{-1}\right) \tag{17}$$

Epstein, Stroschein & Pandolf [39]:

$$Mr. = Mw - 0.5 * (1 - 0.01\,L) * (Mw - 15\,L - 850) \tag{18}$$

Mr. = metabolic cost of running, Mw = metabolic cost of walking, L = clothing weight
   With the maximal error of estimation in the range of running speeds 8–16 km.h.$^{-1}$ about 10%.
   For walking in the range of intensities the oxygen consumption inaccuracy at the speeds from 4 to 10 km.h^{-1} is around 12% [12, 34].
   Running has a greater energy cost than walking on both the track and treadmill. For running, the Léger equation, ACSM [35], and Bunc [12] prediction model appear to be the most suitable for the prediction of running energy expenditure.

The ACSM [35], Pandolf, Givoni & Goldman [37], and Bunc [12] linear prediction equation also closely predict walking energy expenditure, whereas McArdle's [32] table or the equations by Epstein and van der Walt were not as strong predictors of energy expenditure.

For movement speeds lower than 7 km.h^{-1}, the energy cost of running is higher than walking For movement speeds higher than 7 km.h^{-1}, the energy cost of walking is higher and increases exponentially with increasing movement intensity [34] that ACSM [35], Bunc [12] and Léger [38] predictive energy performance models for running straight are more accurate in a young healthy population. For horizontal walking, the ACSM [34, 35], and Pandolfova [37] reduction models also appear to be more accurate than other prediction equations.

The energy intensity of both running and walking represents the total energy consumption using many different mechanisms in the body, including muscle dynamics, blood circulation, and aerobic processes of energy release. In both running and walking energy-intensive experiments in humans, this is usually determined from oxygen consumption and carbon dioxide production values minus the basal metabolic rate at rest to achieve net metabolic performance. The energy intensity of exercise is commonly expressed in two different ways: energy consumed per unit time (metabolic rate or power) or energy consumed per unit distance [40].

The negative relationship between maximal oxygen consumption expressed relative to body mass and coefficient energy cost of running c means that athletes with higher aerobic capacity, higher VO_{2max} have lower values of c, i.e. better running economy [2, 41]. These findings may bet the results of the prolonged duration of the competitions and, thus, of the training performance of these athletes when they are forced to turn out a highly economical performance over a prolonged period of time, and it may also bet the result of a high degree of adaptation to running [25].

4. Energy cost of walking

An energetic economy has been shown to have a large influence on human walking behavior. For example, at a given speed, humans tend to walk with a preferred step length that coincides with minimum metabolic cost [40]. Despite the complexity of the relationship between walking biomechanics and its energy expenditure, relevant studies have shown significant contributors to overall walking energy intensity, such as step-by-step work to redirect the center of gravity and energy in generating muscle strength to support weight transfer and swing. Feet.

The biomechanics of complex movements, such as those that occur when walking and running, which involve a large number of cooperating body segments, can be better understood by considering the energy counterpart, ie the energy expended on muscle contraction, which must work continuously to drive the body forward. Running and especially walking are basic physical activities to which a person is maximally adapted. This adaptation has evolved over many generations in order to minimize the energy requirements of a given physical activity. Walking is an energetically beneficial physical activity, its energy intensity is only about 50% higher than the basal metabolic rate (at speed of 0.6 ms^{-1} it is about 2.44 W.kg^{-1}) [16], and this has in the past allowed populations to expand their ecological niches. Conversely, running can be very challenging and can be continued without slowing down for untrained individuals for less than an hour and sprinting for a much

Dependence of c on speed of movement

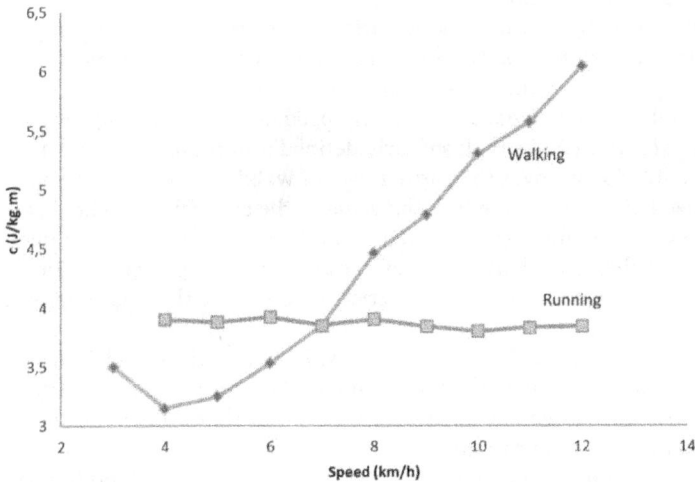

Figure 1.
Dependence of energy cost coefficient on speed of movement in walking and running.

shorter time; but while the energy intensity of walking varies with the speed of movement, when running the same distance, the energy expended, although higher overall, is independent of movement speed [32].

Our older study [34] tries to answer the question of the energy cost of walking (VO2) could be accurately predicted with the simple models which analyze the relationship oxygen uptake-speed of walking. Employing the new modification of this model from 1986 [42] to analyze VO_2 - speed of walking relationship leads to the elaboration of a simple linear model, two-compartment linear model, a polynomial model of second-order and monoexponential model of the metabolic cost of treadmill walking. To verify and compare these models 87 males, age ranged from 19 to 62 years, were evaluated on a motor-driven treadmill. They walked at 0% slope on a treadmill at various velocities ranged from 3 to 12 $km.h^{-1}$.

The linear model has in range of intensities 3–12 $km.h^{-1}$ a form of $VO_2.kg^{-1}$ $(ml.kg^{-1}.min^{-1}) = 5.228*v$ $(km.h^{-1})$-11.158, r = 0.812, S_{EE} = 4.16 $ml.kg^{-1}.min^{-1}$. The two-compartment linear model has in range of intensities of 3–7 $km.h^{-1}$ a form of $VO_2.kg^{-1} = 3.207*v(km.h^{-1})$-1.777, r = 0.932, and S_{EE} = 1.5 $ml.kg^{-1}.min^{-1}$. In the range of 7.1–12 km. $VO_2.kg^{-1} = 7.120*v$-29.168, r = 0.901, S_{EE} = 3.78 $ml.kg^{-1}.min^{-1}$. In the range of intensities from 3 to 12 $km.h^{-1}$ a polynomial model was found in the form $VO_2.kg^{-1} = 4.501$-0.108*v + 0.379*v^2, r = 0.891, S_{EE} = 4.43 $ml.kg^{-1}.min^{-1}$, and the exponential model had a form $VO_2.kg^{-1} = 4.360*exp.(0.223*v)$, r = 0.861, S_{EE} = 6.84 $ml.kg^{-1}.min^{-1}$. All these correlation coefficients were highly significant (p < 0.001 in all cases) [34].

It was concluded that when applied to adult population, the models provide a reasonable estimate of the actual requirement for treadmill walking provided the subjects in an oxygen uptake steady-state [43]. From the above, an important conclusion for practice follows: with adequate accuracy of about 10%, a linear model of the dependence of oxygen consumption and walking speed can be used in the range of walking speeds of 4–12 $km.h^{-1}$.

As with other researches for $VO_2.step^{-1}$ or speed of movement, we have found U-shaped curves of the coefficient energy cost of walking (see **Figure 1**). The minimum was at a speed of about 4 $km.h^{-1}$. This finding supports the speculation that does exists the "optimal" speed of moving which reflects the minimal energy expenditure during walking [34].

The energy coefficient of walking varies depending on the increasing speed of walking. In contrast, the coefficient c for running is practically constant in the range of running speeds 6–15 km.h^{-1}(see **Figure 1**) [7, 44].

In general, the dependence on walking speed or number of steps has a nonlinear parabolic course with a clearly defined minimum of around 4 km.h^{-1} see Figure [34, 42, 45, 46] over the entire range of walking load intensity intensities. From a speed of about 4.5 km.h^{-1}, the value of the coefficient c when walking increases exponentially. For practical use, on the basis of the above, a linear description of the dependence between the energy or oxygen consumption and the speed of movement can be used in practice to determine the energy intensity of the movement.

For movement speeds lower than approx. 7 km.h^{-1}, the coefficient of energy intensity of walking is lower than for running [34]. For practice, this means that in the case of mainly patients, walking at speeds lower than 7 km.s^{-1} is more energetically advantageous than running at the same speed.

During treadmill running most well-trained runners run at step frequencies that minimize their energy expenditure. However, outdoor running, with air resistance and wind, is different from treadmill running [23, 47].

5. Air resistance by movement

The resistance of the environment, in our case walking and running air, is characterized by forces that act against the movement of an object that moves in space. These resistive forces act in the opposite direction to the speed of the approaching flow, thus slowing down the object. Unlike other resistance forces, resistance depends directly on speed, because it is a component of the net aerodynamic force acting against the direction of movement, on the front profile of the moving individual, and on the air density. Therefore, world records on sprinters have often been broken at higher altitudes, where the air density is lower [48, 49].

Air resistance, or drag, can be put into one of three categories; lift induced, parasitic, and wave. Each of these types of air resistance affects an object's ability to stay up and the power it will need to keep it there [49].

Lift-induced air resistance happens as the result of the creation of lift on a three-dimensional lifting body (wing or fuselage).

Parasitic drag happens when a solid object moves through a fluid. This type of air resistance is made up of lots of components like "form drag" and "skin friction drag".

Wave drag is made when an object moves at a high speed through a compressible fluid.

Air resistance is usually calculated using the "drag equation", which determines the force experienced by an object moving through a fluid or gas at a relatively large velocity. This can be expressed mathematically as [49, 50]:

$$F_D = 0.5 * \rho * v^2 * C_D * A \qquad (19)$$

In this equation, FD represents the resistance force, ρ is the air density, v is the velocity of the object relative to the speed of sound, A is the cross-sectional area, and CD is the coefficient of resistance. The result is what is called "quadratic resistance." For movement in an air environment, these constants can be determined as follows.

$$A = 0.266 * (0.2025 * \text{height}^{0.725} * \text{body mass}^{0.425}) \text{ [21]}, \ C_D = 0.9 \text{ [51]} \qquad (20)$$

The energy required to overcome air resistance was estimated at 2% for running outdoors at 5 m.s on a calm windless day [48]. Jones and Doust showed that HR was about 3–4 beats higher when running outdoors (quiet day) compared to running on a treadmill [52]. Pugh found an increase in VO_2 of about 14% as the rate of the ventilator in the laboratory increased from 0 to 10 m.s^{-1} for a subject running on a treadmill at 3.75 m.s^{-1} [23]. This would correspond to the difference between a treadmill running at 3.5 m.s^{-1} and running at the same speed at a headwind of 6.5 m.s-1. Based on our study of trained runners, there is a significant difference in running energy intensity and at running speeds higher than 13 km.h^{-1} [10, 12]. In another study, 6 runners ran in the headwind at a speed of about 6.5 ms^{-1}. The HR value was 4–8 bpm.min^{-1} higher in the headwind compared to the windless, which illustrates the significant effects that wind can have on the energy intensity of the run [53].

The technique of movement changes depending on the increasing resistance of the air at higher speeds. Therefore, in order to maintain the correct movement technique even at speeds above 13 km.h^{-1} running behind the car is often used in the training of runners, which reduces the direct impact of air resistance on the runners.

6. Movement programs based on walking and/or running

Walking and running are very often used in intervention programs for cultivating fitness or for body mass reduction [3, 29, 54, 55]. Exercise intervention with a mean weekly energy intensity of 20.40 ± 4.51 kcal.kg^{-1}.week^{-1} or with a mean energy cost per day of 5.4 kcal.kg^{-1}.day^{-1} which are applied for at least 7 weeks, will cause significant changes in functional and morphological parameters. These changes are independent of age and gender. In the case of weight, the total energy intensity of physical activities increases with increasing body mass [3, 27].

Physical intervention based on walking or running with an intensity corresponding to 80–90% SF_{max}, at least 80% of the total load must consist of running or walking for at least 8 weeks will cause changes in aerobic capacity expressed by changes in VO_{2peak} are on average around 16% of the initial value. We find the same relative change in the speed of movement at which the load on the treadmill is terminated due to subjective exhaustion. The weight reduction is around 14% of the initial weight and the average improvement of the kinetic load assumptions as measured by the ECM / BCM coefficient is around 15% [3, 54].

Recent cross-sectional studies have demonstrated the ability to economize movement, either alone or in combination with V02max, a crucial factor that may explain a substantial portion of performance variations between trained long-distance runners and untrained subjects of comparable levels of exercise and fitness. Limited data from short-term and long-term longitudinal research also suggest that the success of endurance running is related to training and improving the economics of movement leading to a reduction in the energy intensity of movement [53].

In practice, this leads to the clear conclusion that the first step in any endurance training is to improve the economics of running - running techniques leading to reduced energy consumption and delayed fatigue due to depletion of energy resources stored in the body.

7. Discussion

Walking and running are the basic locomotor activities of a person. They are not demanding on the environment and are implemented in practically any weather and in almost all environments - on flat and changing surfaces, movement on the plane, and downhill or uphill. We adopt very well to these forms of physical activity, which results from their long-term use for livelihood and the implementation of work and leisure activities. The energy intensity coefficient of walking depends on the speed of movement and reaches a minimum at a speed of about 4 km.h^{-1} [34]. On the contrary, the coefficient of energy intensity of the run is practically independent of the running speed. The minimum dependence of the coefficient of walking energy intensity on its speed is probably due to the optimal use of the recovery of elastic forces at this speed of movement.

This minimum of energy intensity is often used in the rehabilitation of cardiac patients because the changes caused by the speed corresponding to this minimum are the smallest [56, 57].

Evaluation of the degree of adaptation to running, with the help of c coefficient as an additional characteristic during laboratory tests, enables us to ascertain, along with other parameters, not only the effectiveness of training procedures, but also helps in the evaluation of the technique of the movement performed. This is essential in sports events where training is started at an early age and enables us to determine the energy cost of the training stimulus used [1, 2, 24].

Movement economization in the case of long-term exercise loads is associated with the maximum use of automated movements, which are less energy-intensive than non-automated movements. In practice, this makes key recommendations. At the beginning of each movement intervention, it is always necessary to focus on economizing movement, improving technique \ movement, and only then concentrating on managing the required volume of exercise loads [41].

To master the necessary movement techniques in the case of deepening fatigue, ie in the case of deepening acidosis and reducing the amount of energy substrates is possible only as a result of long-term intensive training [4, 32]. Running economics - the energy intensity of running is primarily dependent on completed training, but the genetic disposition of the runner also plays a role here, i.e. its current level is given by the intersection of genetic preconditions and completed running training [58].

Based on the energy cost of walking or running for a particular individual, it is possible to design a movement intervention that allows you to optimize the effect of this intervention and mainly minimize the time required for this intervention [59].

Evidence suggests that several internal (sex, running biomechanics, anatomy) and external factors (experience, mileage, training routines) may contribute to the risk of injury in recreational runners [60]. Good exercise technique, its good economy, a good value of the energy intensity coefficient of running c, can significantly delay the onset of fatigue during long-term running training load and can act preventively against muscle injuries [61]. Therefore, special attention should be paid to cultivating running techniques in preparation for long-distance races, such as marathons and ultra-endurance races, in order to ensure the necessary condition without increasing the risk of injury from overload.

8. Conclusions

Evaluation of energy cost of majority of physical movement activities and subsequent cultivation, used to influence fitness or in primary and secondary prevention, allows to increase the effectiveness of the applied physical intervention. At the same time, it can delay the onset of fatigue and thus reduce the incidence of muscle injuries. Assessing the energy intensity of running or walking in laboratory or field functional tests can significantly expand the information content of these surveys and should therefore be an integral part of these surveys. Models relating energy and intensity to exercise are useful for quantifying the training load of both recreational and trained runners and allow you to minimize the time devoted to endurance training.

Author details

Vaclav Bunc
Faculty of P.E. and Sports Charles University Prague, Czech Republic

*Address all correspondence to: bunc@ftvs.cuni.cz

IntechOpen

References

[1] Astrand PO, Rodahl K. Textbook of Work Physiology. New York, USA: McGGraw Hill; 1986

[2] Bunc V, Heller J. Energy cost of running in similarly trained men and women. European Journal of Applied Physiology and Occupational Physiology. 1989;**59**(3):178-183. DOI: 10.1007/BF02386184

[3] Bunc V. Obesity - causes and remedies. Physical Activity Review. 2016;**4**:50-56. DOI: 10.16926/par.2016.04.06

[4] Bunc V. Energy cost of treadmill running in non-trained females differing in body fat. The Journal of Sports Medicine and Physical Fitness. 2000;**40**(4):290-296

[5] Vernillo G, Giandolini M, Edwards WB, Morin JB, Samozino P, Harvais N, et al. Biomechanics and physiology of uphill and downhill running. Sports Medicine. 2017;**47**:615-629. DOI: 10.1007/s40279-016-0605-y

[6] di Prampero PE, Atchou G, Brückner JC, Moia C. The energetics of endurance running. European Journal of Applied Physiology and Occupational Physiology. 1986;**55**(3):259-266. DOI: 10.1007/BF02343797

[7] Margaria R. Biomechanics and Energetics of Muscular Exercise. Oxford, UK: Claredon; 1976

[8] Zamparo P, Matteo Cortesi M, Gatta G. The energy cost of swimming and its determinants. European Journal of Applied Physiology. 2020;**120**(1): 41-66. DOI: 10.1007/s00421-019-04270-y

[9] Bunc V, Heller J. Energy cost of running in young and adult female athletes. Ergonomics. 1994;**37**(1):167-174. DOI: 10.1080/00140139408963635

[10] Bunc V, Heller J. Exercise intensity coversion from a bicycle ergometer to a treadmill. The Journal of Sports Medicine and Physical Fitness. 1991; **Fitness 31**:490-493

[11] di Prampero PE. The energy cost of human locomotion on land and in water. International Journal of Sports Medicine. 1986;**7**(2):55-72. DOI: 10.1055/s-2008-1025736

[12] Bunc V, Heller J, Leso J. Energy requirements of running on a treadmill and in the field. Czech Republic Medical Journal. 1986;**125**:1391-1394

[13] Morgan DW, Martin PE, Krahenbuhl GS. Factors affecting running economy. Sports Medicine. 1989;**7**:310-330

[14] Bunc V, Heller J. Energy demands in women's running on the treadmill and cross county running. Czech Republic Medical Journal. 1990;**129**:650-653

[15] de Ruiter CJ, van Daal S, van Dieën JH. Individual optimal step frequency during outdoor running. European Journal of Sport Sciences. 2020;**20**(2): 182-190. DOI: 10.1080/17461391.2019. 1626911

[16] Hall C, Figueroa A, Ferrnhall BO, Kanaley JA. Energy expenditure of walking and running: Comparison with prediction equations. Medicine & Science in Sports & Exercise. 2004;**36** (12):2128-2134. DOI: 10.1249/01. MSS.0000147584.87788.0E

[17] Anderson T. Biomechanics and running economy. Sports Medicine. 1996;**22**(2):76-89. DOI: 10.2165/ 00007256-199622020-00003

[18] Peyré-Tartaruga LA, Dewolf AH, di Prampero PE, Fábrica G, Malatesta D, Minetti AE, et al. Mechanical work as a

(key) determinant of energy cost in human locomotion: Recent findings and future directions. Experimental Physiology. 2021;**106**(9):1897-1908. DOI: 10.1113/EP089313

[19] Black MI, Allen SJ, Forrester SE, Folland JP. The anthropometry of economical running. Medicine & Science in Sports & Exercise. 2020; **52**(3):762-770. DOI: 10.1249/MSS.0000 000000002158

[20] Lacour JR, Bourdin M. Factors affecting the energy cost of level running at submaximal speed. European Journal of Applied Physiology. 2015;**115** (4):651-673. DOI: 10.1007/s00421-015-3115-y

[21] Snyder KL, Farley CT. Energetically optimal stride frequency in running: The effects of incline and decline. The Journal of Experimental Biology. 2011; **214**:2089-2095

[22] Arsac LM, Locatelli E. Modeling the energetics of 100-m running by using speed curves of world champions. Journal of Applied Physiology. 2002;**92** (5):1781-1788. Crossref. Retrieved from: https://www.ncbi.nlm.nih.gov/pubmed/11960924

[23] Pugh LGCE. The influence of the wind resistence in running and walking and the mechanical efficiency of work against horizontal forces. Journal of Physiology (London). 1971;**213**:255-276

[24] Costill DL. A scientific approach to distance running. Track Fields News. 1979;**14**:12-40

[25] Cavagna GA, Mantovani M, Willems PA, Musch G. The resonant step frequency in human running. Pflügers Archiv - European Journal of Physiology. 1997;**434**:678-684

[26] Arellano CJ, Kram R. Partitioning the metabolic cost of human running: A task-by-task approach. Integrative and Comparative Biology. 2014;**A54**(6): 1084-1098. DOI: 10.1093/icb/icu033

[27] Cebolla i Martí A, Álvarez-Pitti JC, Guixeres PJ, Lisón JF, Baños RR. Alternative options for prescribing physical activity among obese children and adolescents: Brisk walking supported by an exergaming platform. Nutrición Hospitalaria. 2014;**31**(2):841-848. DOI: 10.3305/nh.2015.31.2.7929

[28] Walker JL, Murray TD, Jackson AS, Morrow JR Jr, Michaud TJ. The energy cost of horizontal walking and running in adolescents. Medicine and Science in Sports and Exercise. 1999;**31**(2):311-322

[29] Murtagh EM, Murphy MH, Boone-Heinonen J. Walking: The first steps in cardiovascular disease prevention. Current Opinion in Cardiology. 2010; **25**(5):490-496. DOI: 10.1097/HCO.0b013e32833ce972

[30] Cavagna GA, Saibene FP, Margaria R. Mechanical work in running. Journal of Applied Physiology. 1964;**19**:249-256

[31] Giandolini M, Pavailler S, Samozino P, Morin P, Horvais N. Foot strike pattern and impact continuous measurements during a trail running race: Proof of concept in a world-class athlete. Footwear Science. 2015;**7**(2): 127-137. DOI: 10.1080/19424280.2015.1026944

[32] Gianluca V, Giandolini M, Edwards WB, Morin JB, Samozino P, Horvais N, et al. Biomechanics and physiology of uphill and downhill running. Sports Medicine. 2017;**47**:615-629. DOI: 10.1007/s40279-016-0605-y

[33] McArdle W, Katch F, Katch V. Exercise Physiology: Energy Nutrition, and Human Performance. Baltimore, MD: Williams & Wilkins; 2001

[34] Bunc V, Dlouhá R. Energy cost of treadmill walking. The Journal of Sports

Medicine and Physical Fitness. 1997;37(2):103-109

[35] ACSM. American College of Sports Medicine. Guidelines for Exercise Testing and Prescription. 6th ed. Baltimore, MD: Lippincott Williams & Wilkins; 2000. pp. 163-176

[36] van der Walt W, Wyndham C. An equation for prediction of energy expenditure of walking and running. Journal of Applied Physiology. 1973;34: 559-563

[37] Pandolf K, Givoni B, Goldman R. Predicting energy expenditure with loads while standing or walking very slowly. Journal of Applied Physiology. 1978;43:577-581

[38] Léger L, Mercier D. Gross energy cost of horizontal treadmill and track running. Sports Medicine. 1984;1: 270-277

[39] Epstein Y, Stroschein LA, Pandolf KB. Predicting metabolic cost of running with and without backpack loads. European Journal of Applied Physiology and Occupational Physiology. 1987;5(6):495-500

[40] Faraji S, Wu AR, Ijspeert AJ. A simple model of mechanical effects to estimate metabolic cost of human walking. Scientific Reports. 2018;8(1):10998. DOI: 10.1038/s41598-018-29429-z

[41] Saunders PU, Pyne DB, Telford RD, Hawley JA. Factors affecting running economy in trained distance runners. Sports Medicine. 2004;34:465-485

[42] Workman JM, Armstrong BW. Metabolic cost of walking: Equation and model. Journal of Applied Physiology. 1986;61(4):1369-1374. DOI: 10.1152/ jappl.1986.61.4.1369

[43] Donelan JM, Kram R, Kuo AD. Mechanical work for step-to-step transitions is a major determinant of the metabolic cost of human walking. The Journal of Experimental Biology. 2002; 205:3717-3727

[44] Falls HB, Humphrey LD. Energy cost of running and walking in young women. Medicine and Science in Sports. 1976;8:9-13

[45] Grabowski A, Farley CT, Kram R. Independent metabolic costs of supporting body weight and accelerating body mass during walking. Journal of Applied Physiology. 2005;98:579-583. DOI: 10.1152/japplphysiol.00734

[46] Pivarnik JM, Sherman NW. Responses of aerobically fit men and women to uphill/downhill walking and slow jogging. Medicine and Science in Sports and Exercise. 1990;22:127-130

[47] Pugh LGCE. Oxygen uptake in track and treadmill running with observationon the effect of the air resistence. Journal of Physiology (London). 1970;207:823-835

[48] Davies CTM. Effect of wind assistance and resistence on the forward motion of a runner. Journal of Applied Physiology. 1980;48:702-709

[49] Falkovich G. Fluid Mechanics (a Short Course for Physicists). UK: Cambridge University Press; 2011. ISBN 978-1-107-00575-4

[50] Mero A, Komi PV, Gregor RJ. Biomechanics of sprint running. A review. Sports Medicine. 1992;13(6): 376-392. DOI: 10.2165/00007256-199213060-00002

[51] van Ingen Schenau GJ, Jacobs R, de Koning JJ. Can cycle power predict sprint running performance? European Journal of Applied Physiology and Occupational Physiology. 1991;63(3–4): 255-260. Retrieved from: https://www. ncbi.nlm.nih.gov/pubmed/1761017. Crossref

[52] Jones AM, Doust JH. A 1% treadmill grade most accurately reflects the energetic cost of outdoor running. Journal of Sports Sciences. 1996;**14**(4):321-327. DOI: 10.1080/02640419608727717

[53] de Ruiter CJ, van Daal S, van Dieën JH. Individual optimal step frequency during outdoor running. European Journal of Sport Science. 2020;**20**(2): 182-190. DOI: 10.1080/17461391.2019. 1626911

[54] Bunc V, Hrasky P, Skalska M. Possibilities of body composition influence by walking program in senior women. Journal of Aging and Physical Activity. 2012;**20**:S289-S290

[55] Fanning J, Opina MT, Leng I, Lyles MF, Nicklas BJ, Rejeski WJ. Empowered with movement to prevent Obesity & Weight Regain (EMPOWER): Design and methods. Contemporary Clinical Trials. 2018;**72**:35-42. DOI: 10.1016/ j.cct.2018.07.010

[56] Girold S, Rousseau J, Le Gal M, Coudeyre E, Le Henaff J. Nordic walking versus walking without poles for rehabilitation with cardiovascular disease: Randomized controlled trial. Annals of Physical and Rehabilitation Medicine. 2017;**60**(4):223-229. DOI: 10.1016/j.rehab.2016.12.004

[57] Han P, Zhang W, Kang L, Ma Y, Fu L, Jia L, et al. Clinical evidence of exercise benefits for stroke. Advances in Experimental Medicine and Biology. 2017;**1000**:131-151. DOI: 10.1007/978-981-10-4304-8_9

[58] Boullosa D, Esteve-Lanao J, Casado A, Peyré-Tartaruga LA, Gomes da Rosa R, Del Coso J. Factors affecting training and physical performance in recreational endurance runners. Sports (Basel). 2020;**8**(3):35. DOI: 10.3390/ sports8030035

[59] Hausswirth C, Lehénaff D. Physiological demands of running during long distance runs and triathlons. Sports Medicine. 2001;**31**(9):679-689. DOI: 10.2165/00007256-200131090-00004

[60] Malisoux L, Delattre N, Urhausen A, Theisen D. Shoe cushioning influences the running injury risk according to body mass: A randomized controlled trial involving 848 recreational runners. The American Journal of Sports Medicine. 2019;**48**: 473-480

[61] Pons V, Riera CX, Martorell M, Sureda A, Tur JA, Drobnic F, et al. Calorie restriction regime enhances physical performance of trained athletes. Journal of the International Society of Sports Nutrition. 2018;**15**:12

Chapter 10

Mechanical Limits of Cardiac Output at Maximal Aerobic Exercise

Sheldon Magder

Abstract

This chapter uses an analytic approach to the factors limiting maximal aerobic exercise. A person's maximal aerobic work is determined by their maximal oxygen consumption (VO$_2$max). Cardiac output is the dominant determinant of VO$_2$ and thus the primary determinant of population differences in VO$_2$max. Furthermore, cardiac output is the product of heart rate and stroke volume and maximum heart rate is determined solely by a person's age. Thus, maximum stroke volume is the major factor for physiological differences in aerobic performance. Stroke output must be matched by stroke volume return, which is determined by the mechanical properties of the systemic circulation. These are primarily the compliances of each vascular region and the resistances between them. I first discuss the physiological principles controlling cardiac output and venous return. Emphasis is placed on the importance of the distribution of blood flow between the parallel compliances of muscle and splanchnic beds as described by August Krogh in 1912. I then present observations from a computational modeling study on the mechanical factors that must change to reach known maximum cardiac outputs during aerobic exercise. A key element that comes out of the analysis is the role of the muscle pump in achieving high cardiac outputs.

Keywords: aerobic limit, oxygen consumption, stressed volume, venous return, cardiac output, stroke volume, heart rate, time constants

1. Introduction

Sustainable work at high levels of energy consumption requires oxygen (O$_2$) based metabolism. Maximum O$_2$ consumption (VO$_2$) thus determines a person's maximal aerobic power [1]. A young active healthy male of standard size can increase VO$_2$ from a resting value of around 0.25 L/min to between 3.00 and 3.5 L/min, a 12–14 fold increase [1, 2]. In elite athletes, values greater than 6.0 L/min have been measured [3]. The physiological basis of these numbers can be understood by considering the Fick principle, which is essentially a statement of the conservation of mass [1, 4]. VO$_2$ is the product of how much volume per minute/min (L/min) is delivered to tissues, in other words, cardiac output, and how much O$_2$ is extracted from each volume unit of blood [1, 2].

$$\dot{V}O_2 = Q \, x \left[Hb \right] x 1.36 \left(Sat_a O_2 - Sat_v O_2 \right) \tag{1}$$

Q is cardiac output (L/min), [Hb] is hemoglobin concentration (in g/L), 1.36 is the constant for the amount of O_2 (ml) per g of Hb, Sat_aO_2 is the arterial O_2 saturation (as a decimal) and Sat_vO_2 is the venous O_2 saturation (as a decimal). Thus, the limit of aerobic function is based on the maximum extraction of O_2 from the blood and the maximum cardiac output.

The capacity of arterial blood to carry O_2 is determined by the concentration of hemoglobin ([Hb]) and the amount of O_2 that each gram of Hb can bind [5]. The constant for binding of O_2 to Hb with no other substances present is 1.39 ml O_2 per gram of Hb, but normally other molecules in blood, such as methemoglobin and carboxyhemoglobin, take up some of the binding sites. Thus, constants of 1.34–1.39 are used in the literature to account for these factors. The actual content of O_2 in blood is dependent upon [Hb] and the saturation of Hb molecules with O_2; the saturation is in-turn is dependent upon the partial pressure of O_2 in blood (PO_2 in mmHg). [Hb] concentration thus sets the upper limit of how much O_2 is present to be extracted from the blood. As an example, with a [Hb] of 145 g/L, a saturation of 98%, and capacity of Hb to carry O_2 of 1.36 ml O_2/g Hb, the arterial O_2 content would be 197.1 ml/L. The saturation of arterial blood usually is slightly less than 100% because of some shunting of blood across the lungs and venous blood returning to the left ventricle from the coronary circulation. [Hb] is similar in a standard male and endurance athlete (unless there has been some kind of unfair manipulation of [Hb]!) so that this factor does not play a large role in differences in maximum VO_2. During exercise [Hb] increases slightly because of a loss of plasma and hemoconcentration [6].

The O_2 content of blood returning to the right heart gives the overall extraction of O_2 by all tissues. This is called mixed venous O_2 content (M_vO_2). At rest, about 25% of the arterial blood O_2 content is extracted, which gives a M_vO_2 of around 150 ml/L in both a standard male and elite athlete [6]. During peak aerobic exercise, the greatest proportion of the blood goes to the working muscle, which is capable of extracting almost all the delivered O_2 it receives at peak performance. Under resting conditions about 60% of blood flow goes to the muscle vasculature and 40%, or about 2 L/min, goes to the non-muscle vasculature [7, 8]. At peak exercise, the amount going to non-working muscle remains largely unchanged, or decreases by a small amount, so that greater than 90% of blood flow goes to the working muscle [7], which at peak performance can extract almost all the O_2 it receives. The percent of blood flow going to non-working tissues sets a lower limit of O_2 extraction [9]. The maximal amount of O_2 extracted is similar in standard young healthy males and endurance athletes, although extraction can be slightly greater in endurance athletes. This is likely because they have larger amount of muscle mass per total body mass, and thus a higher fraction of blood flow can go to the working muscle, which results in greater total extraction. A greater capacity to endure discomfort may also play a role. A typical M_vO_2 at peak performance is in the range of 22 ml/L in the standard male and 18 ml/L in elite endurance athletes, which is less than a 1% difference in the total amount extracted [10]. Thus, differences in O_2 extraction between standard and elite athletes contribute little to differences in their maximum VO_2 unless the arterial O_2 carrying capacity is significantly increased, although this potentially could limit extraction by increasing blood viscosity and reducing blood flow to tissues.

Based on the Fick equation, the other determinant of maximum VO_2 is cardiac output. In a typical young male, cardiac output can increase from a resting value of around 5 L/min to 20–25 L/min, a 4–5 fold increase [1, 2]. In elite athletes maximal cardiac output can be in the range of 30 L/min to even over 40 L/min in some high performing cyclist and cross-country skiers [11]. The athletes thus can have a 6–7 fold increase in cardiac output from resting levels and this increase in cardiac output is the major factor explaining their higher aerobic capacity [4]. If [Hb] is normal there is a tight linear relationship between cardiac output and VO_2 that is independent of body size, fitness, or age [2, 4]. The slope of this relationship is the same in women and men but the relationship is shifted downward in women because of they generally have lower [Hb] [2].

Cardiac output is the product of beats per minute, that is, heart rate, and stroke volume. Maximum heart rate at peak aerobic performance is solely determined by age and not by differences in fitness, body size, heart size, or sex; the rise in heart rate is dependent upon the percent of the maximal capacity of the muscles being used [2]. This means that the primary difference in aerobic power of the standard male and elite aerobic athletes is the maximum possible stroke volume for that person [1, 3]. Furthermore, stroke volume is dependent upon heart size, which for healthy hearts is related to lean body size as determined by the person's genetic make-up [12]. There is little change in stroke volume capacity with training [10], although increases in maximum stroke volume often are observed in studies with training [10, 13]. These observed increases in stroke volume are likely related to reductions in submaximal heart rate, which occur due to alterations in neuro-humeral mechanisms with training [9]. A lower heart rate at a given VO_2 requires that there be a larger stroke volume for the same venous return and cardiac output so that the relationship of cardiac output to VO_2 is maintained, but this does not mean that there was an intrinsic change in heart structure.

2. Basic principles of the determinants of blood flow in the circulation

It often is thought that blood flow around the circuit is dependent on the arterial pressure regenerated by the heart [14]. This view is of presented as an electrical model with the arterial pressure being the equivalent of a fixed voltage from a source. In this construct, vascular volume, which the electrons in the circuit, is not a fixed value, but can increase or fall based on current for the fixed pressure drop. In contrast, Arthur Guyton [15], and for that matter, Ernest Starling [16], used a hydraulic approach in which the elastic energy, that is pressure, produced by a fixed volume in the circuit determines the return of blood to the heart. The action of the heart in this approach is to pump the returning blood back to the circuit [17]. In the Guyton approach, blood flow around the circuit is determined by two functions: cardiac function and a function that describes the return of blood to the heart from a large venous compliant region [15]. These are discussed next.

2.1 Cardiac function

The basis of cardiac function is the Frank-Starling law, which says that the greater the initial cardiac muscle length the greater the force produced by the heart up to a limit [16]. The determinants of cardiac output are heart rate and stroke volume, and stroke volume is determined by the preload, afterload and contractility. Cardiac function is plotted with right atrial pressure (Pra) at the end-of diastole. This determines right ventricular end-diastolic muscle length, and the preload, on the x-axis, and cardiac output on the y-axis (**Figure 1**) [18]. This relationship assumes a constant heart rate, afterload and contractility. An increase in cardiac function is produced by an increase in heart rate, increase in contractility, or a decrease in afterload and is indicated by upward shift of the curve (**Figure 1**). The opposites cause a decrease in cardiac function and a

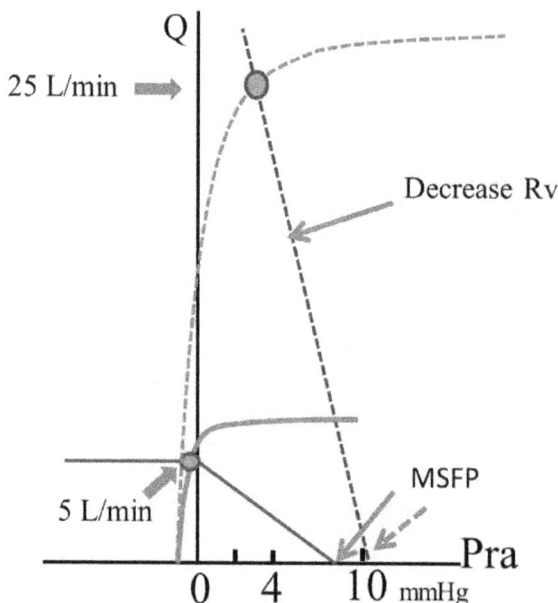

Figure 1.
Schematic plots of venous return and cardiac function curves at rest and maximal aerobic exercise. The resting state is shown with solid lines and exercise state with dashed lines. The change in cardiac output was from 5 to 25 L/min. The x-axis is right atrial pressure (Pra) in mmHg and the y-axis is blood flow in L/min. The curved red lines are cardiac function curves and the blue lines are venous return curves. The slope of the venous return curve is −1/resistance to venous return (Rv). During exercise there is a marked increase in the cardiac function curve due to primarily to the rise in heart rate and to a lesser extent, stroke volume. This is matched by a marked decrease in Rv and a small increase in MSFP due a decrease in vascular capacitance.

downward shift of the curve. Importantly, the cardiac function curve has a sharp plateau [19], and when reached, further increases in preload, that is Pra, do not increase cardiac output.

2.2 Venous return function

The typical total blood volume of a 70 kg male is approximately 5.5 L. When there is no flow in the circulation the contained volume still stretches vascular walls and creates a pressure of 7–10 mmHg; this is called mean circulatory filling pressure (MCFP) [15, 20]. About 70% of vascular volume resides in small veins and venules. The compliance of the walls of these vessels, that is change in volume per change in pressure, is 30–40 times greater than that of arterial and capillary vessels and seven times the compliance of the pulmonary vessels [21]; because systemic veins dominate the volume of the circulation, they are the major determinant of MCFP [22]. Under flow conditions, volume redistributes throughout the vasculature so that there can be some change in volume and pressure in the veins and venules. Since the pressure in veins and venules is the upstream pressure determining blood return to the heart, it is given a separate name, mean systemic filling pressure (MSFP). When there is no flow, MSFP and MCFP are equal as is Pra.

Another important concept is vascular capacitance [23, 24]. Some of the total vascular volume just rounds out vessels, but this volume does not stretch vessel walls. This volume is thus "unstressed" in that it does not produce a pressure. Only the volume above unstressed volume stretches vessels walls and accordingly is called stressed volume. Under resting conditions about 70% of blood volume is unstressed and 30%, or about 1.3–1.4 L, is stressed [25]. Only this stressed volume produces the elastic force that drives venous blood back to the heart. However, activation of sympathetic activity can result in constriction of veins and venules and thereby convert unstressed into stressed volume [24, 26–28]. By this means, unstressed volume acts as a vascular reserve, and stressed volume can be increased by 6–10 ml/kg, and sometimes more. This is the equivalent to a vascular infusion of fluid and can greatly increase MSFP almost instantly.

Venous blood returns to the right heart through the resistance downstream from the venous compliant regions [15]. Although small, and only produces a pressure drop only of 4–8 mmHg, this resistance it is a major determinant of the return of blood to the heart because it controls the emptying of the large upstream venous reservoir.

The final determinant of venous return is Pra. This is the downstream pressure for venous drainage. In this context, it can be considered that the primary role of the heart in the control of cardiac output is to keep Pra low so as to allow venous return [18]. In this sense, the right ventricle has primarily a "permissive function" in that it "allows" venous blood to return. Importantly, the best that the heart can do is lower Pra to zero, that is, atmospheric pressure. Below this value the pressure inside the floppy great veins is less than the surrounding pressure and collapse, thereby creating flow limitation or what is called a vascular waterfall. When this happens, lowering Pra further does not increase venous return and accordingly, cardiac output [29].

Venous return was plotted by Arthur Guyton with flow on the y-axis and Pra on the x-axis (**Figure 1**) [15]. The x-intercept is MSFP and the slope of the line is the negative inverse of the resistance to venous return (−1/Rv). The venous return curve has the same axes as the cardiac function curve. Thus cardiac function and return function can be plotted on the same graph and this plot can be used to mathematically solve the interaction of these two functions [15].

2.3 Interaction of the return function and cardiac function

Actual cardiac output is determined by the intersection of the cardiac function and return function (**Figure 1**) [15]. An isolated increase in cardiac function produces an increase in cardiac output and a decrease in Pra. An isolated increase in the return function produces an increase in cardiac output with a rise in Pra [18]. A study in normal young males showed that at the onset of pedaling on an upright stationary cycle, Pra immediately increased from –2 to ~4 mmHg [30] and then changed little all the way to maximum effort [31]. This indicates that after an initial moderate increase in preload because of volume being squeezed out the working muscle, increases in cardiac and return functions are perfectly matched. Interestingly, in a group of patients with denervated transplanted hearts, although it took a longer time for equilibration to occur, as exercise continued they too had little change in Pra with increasing cardiac outputs. The slope of the rise in their cardiac output with the rise in VO_2 also was the same range as normal subjects [31–33].

2.4 Two compartment model

So far in this discussion, I have applied the Guyton model of the circulation which considers that there is only one large venous compliant region [20]. In 1912 August Krogh [34] observed that if a closed circuit has two regions with different compliances in parallel, changes in the fractional distribution of flow between the two regions produced by changes in their inflow resistances, alters the rate of flow around the system (**Figure 2**). This is because when more volume goes to the less compliant region, the pressure rises in this region, which then increases the rate of

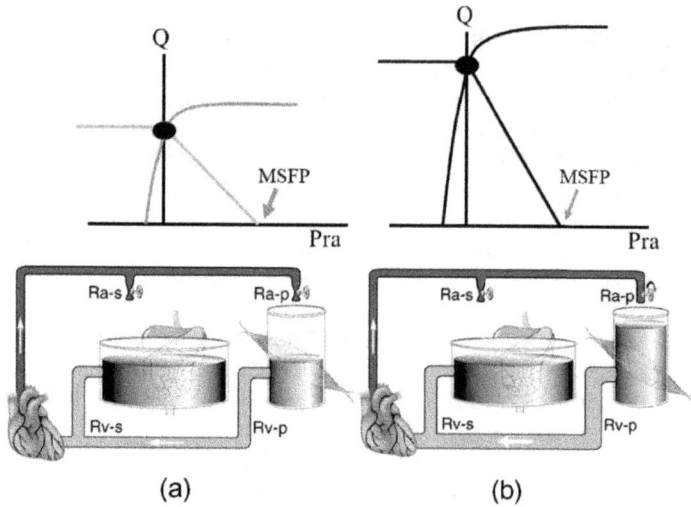

Figure 2.
*The two compartment model of the circulation—Krogh model. In this model there are two parallel venous compartments. One, the equivalent of the splanchnic bed (s) has a large compliance (i.e., large volume for given height) and the other, the equivalent of the peripheral-muscle bed (p), a low compliance (smaller volume for a given height). Flow into each compartment is determined by their arterial resistances (Ra) which act like taps controlling the flows. Drainage occurs through their venous resistances (Rv) which also can be regulated. The top shows the cardiac function-venous return plot as in **Figure 1**. a is the system at rest and b shows what could happen with exercise. Increasing flow to the muscle bed by lowering its inflow reisistance (Ra-p), raises the pressure in this region and increases the outflow. This is seen on the cardiac function-venous return plot as a steeper slope to the venous return curve. This allows an increase in cardiac function to produce a higher cardiac output.*

outflow from that region. This concept was subsequently further developed for the cardiovascular system by Permutt and co-workers [35, 36]. The venous compartment of the splanchnic circulation is much more compliant than that of the muscle vasculature [37, 38]. This makes sense from an evolutionary point of view because there is a lot more space in the abdominal region to take up volume reserves than in the actively contracting and thinner limbs. A shift of blood flow to muscle beds because of metabolic dilation with exercise thus increases net venous return. This appears on the cardiac-venous return plot as a steepening of the slope of the venous return function, which has resistance units. However, the x-intercept, which is MSFP, does not change because when flow is zero the pressures is the same everywhere in the vasculature (**Figure 2**).

Permutt and coworkers further explored the mathematical basis for this. The product of a resistance and compliance draining a vascular bed gives its time constant of drainage (τ) [35, 36]. The τ is the time it takes to get to ~63% of a new steady state pressure and volume when there is a step change in flow into a region. Based on animal studies, estimates of the τ draining the splanchnic bed are in the 20–24 second range, and those of the muscle compartment 4–6 second.

2.5 Importance of τ in the circulation

The τ for filling and draining regions in the vasculature are of great importance because pressure and flow in the system are pulsatile. This creates a periodicity that fixes times for flow into and out of vascular regions. As the frequency of cardiac pulsations increase, the τ of a region can limit flow. As a reminder, τ is the product of the compliance and resistance of a system. Since the left ventricle develops a very high elastance during ejection [39], and thus a low compliance, and it pumps into large conductance vessels with low a low input resistance, the τ of emptying of the left ventricle is much shorter than the τ of emptying of systemic veins returning blood to the heart. However, even then, the ventricle does not eject all its volume during systole. Similarly, right ventricular filling does not normally reach the limit of filling during diastole because there is not enough time to do so. There is thus room in the system to respond to faster inflow to the right ventricle, and for the left ventricle to handle the increased volume per minute. However, as heart rate increases, diastole and systole shorten and there is less time for ejection and filling. Ejection is aided by the marked decrease in arterial resistance with exercise which shortens the τ of ejection. On the diastolic side, there needs to be a shortening of diastolic relaxation and a shortening of the period of ejection to allow more time for flow to come back. The increase in the rate of venous return is aided by the decrease in venous resistance through a number of mechanism. These include a decrease in the venous resistance of muscle because of passive dilation from higher flow as well as possible flow-mediated active dilatation and decreased resistance to the venous drainage of the splanchnic bed which is driven by beta-adrenergic activity [28, 40]. As discussed above, another factor is the distribution of blood flow between the splanchnic and muscle beds which is discussed next.

3. Two compartment computational model of the circulation and determination of maximal cardiac output

Based on the rational above, we adapted a computational model of the circulation so that it had two parallel venous compartments [41]. One compartment represented the splanchnic bed and had a τ at rest of 22 sec and the other represented the muscle compartment with a resting τ of 4 sec. Resting parameters used in the model were based on animal and human data and adjusted to give known resting hemodynamic values in humans [2, 6, 42]. These included a resting heart rate of 65 b/min, cardiac output of 5 l/min, and a mean blood pressure of 93 mmHg. Adjustments were then made in circuit parameters as needed to aim for a peak cardiac output in the range of 20–25 L/min and a mean blood pressure of 115 mmHg. Values for resistances and compliances in the model, and their changes with sympathetic activation, were based on animal studies [26–28, 36, 38, 40]. Modeling cardiac parameters turned out to be very challenging. Two critical assumptions were the limit of diastolic filling of the right heart because it sets the maximum stroke volume and second, the constants that determine the shape of the diastolic passive filling curve of the right and left ventricles because these affect diastolic filling pressure and thus the gradient for venous return. Based on values in the literature, we set the limit of a normal right ventricular diastolic volume to 140 ml. With a heart rate of 180 b/min, this gave an upper limit to the maximum possible cardiac output of 25 L/min.

To simulate the maximal exercise condition we first increased the heart rate to 180 b/min. Without any adjustments in the mechanics of the circulation this actually lowered cardiac output. Even when we shortened the systolic ejection time and the τ for diastolic relaxation of the ventricles, in the absence of adjustment other circuit factors, there still was little change in resting cardiac output. This is because blood needed to come back faster. When we increased the distribution of blood flow going to the muscle bed from 60 to 90% as expected during exercise, t cardiac output increased only moderately and arterial pressure fell markedly, which contributed to lower than expected muscle blood flow. Based on a previous study of baroreceptor regulation of the systemic circulation [28], we reduced the resistance draining the splanchnic bed as well as the capacitance of that bed to produce an increase in stressed volume of 6 ml/kg. From studies on the τ of isolated contracting skeletal muscle [43, 44] we lowered the venous resistance of the muscle bed during the simulated exercise condition from 4 to 2 sec. Each of these individually only produced small changes in cardiac output. However, when they were all put together, and arterial resistances were adjusted to obtain the expected arterial pressure, cardiac output was still only around 17 L/min, and much less than what we were aiming for, which was 20–25 L/min. Another consequence was that the venous pressures markedly increased in the veins of muscles, which means that capillary pressures would have been markedly elevated and there would be major capillary leak.

The two compartment model and the τ draining the regions can be used to give a mathematical formulations of the maximum cardiac output:

$$Q = \frac{\gamma - \left(C_T \; x \; Pra\right)}{F_S \tau_S + F_M \tau_M} \tag{2}$$

Q = cardiac output (L/min), γ = stressed vascular volume (L), C_T = total vascular compliance (ml/mmHg, and is the sum of splanchnic and muscle compliances), F = fractional of total cardiac output to the region, τ = time constant (sec) and subscripts S and M = splanchnic and muscle vasculatures, respectively. This simplification omits the small variation that would occur if volume accumulates or is lost in the pulmonary circuit.

3.1 Muscle pump

The solution for reaching known high values of cardiac output at peak exercise was to add the equivalent of a muscle pump by having active compressions on the muscle venous compartment. This increased cardiac output to the 22–23 L/min range we were aiming for. The muscle pump acted through a number of mechanisms. The marked decrease in arterial resistance that is required to increase blood flow to the working muscle results in capillary and venous pressures that approach arterial pressure. The high venous pressure then stretches the compliant venous walls and a large amount of volume would accumulate in the muscle vasculature. This effectively creates a loss of a large proportion of stressed volume from the circuit. By forcefully squeezing veins during contraction, muscle contractions transiently empty the venous volume to almost zero. This appears as a large venous pulse of flow from the veins of contracting muscles [30, 44, 45]. Transiently lowering local venous volume also ensures that the capillary pressures do not remain persistently high when exercising and thereby reduces capillary leak. Another effect of the muscle pump is that it "speeds" up the movement of blood from the muscle veins to the heart, effectively decreasing the time constant of venous drainage [43]. The importance of this will be discussed below under factors affecting RV filling.

Muscle contractions do not send blood in the retrograde direction because of the presence of a Starling resistor-like mechanism at the arteriolar level [46]. Because of this mechanism, the muscle also does not "suck" blood from the arteries and does not act as an auxiliary pump; rather, it facilitates flow by preventing volume accumulation in the muscle vasculature. When strong enough, muscle contractions likely even pump blood through the tricuspid and pulmonary valves during systole and diastole, and thereby increase the time and the cycles for flow to go through the right heart.

A number of factors determine the effectiveness of the muscle pump. The most obvious is that the force generated must be adequate to compress the venous compartment. In the modeling study, the effect was evident with a force of 0.25 mmHg/ml. The larger the muscle groups, the larger the affect because there is more volume to empty.

4. Implications for peak performance of the observations from the modeling study

Since stroke volume is a key determinant of the maximum cardiac output, a person's innate heart size as determined by their genetic make-up is a key variable. This does not change much with training. Rather, training changes the characteristic of the kinetics of the return. This is largely related to the amount of muscle. Basic muscle mass generally evolves in proportion to heart size and these are genetically linked. Muscle mass, though, can be increased by training and this likely is what accounts for the expected potential 20% increase in aerobic power with training in someone who has not been previously active [10, 13, 47]. More muscle means that more blood can come back to the heart. If someone has already been active aerobically there likely is little more to gain in VO_2-max with training. This does not mean, though, that they cannot do more with what they have by an increase in their anaerobic threshold [48].

The importance of the amount of muscle as a determinant of cardiac output becomes an issue when trying to restore lost muscle. A question that needs to be answered for rehabilitation of lost muscle is what level of aerobic activity is needed to be to promote return of the original muscle mass? If the person's capacity starts very low, it then may not be possible to do a high enough level activity to generate the increase in blood flow in working muscle to generate an increase in aerobic activity. Age likely plays a role, too, in the capacity of muscle to recover. There are many examples of athletes with high aerobic power who have had significant injuries but still are able to return to performing at high aerobic levels. With aging though, return of full mass might be less effective despite high levels of training. There is evidence of this in recovering critically ill patients beyond age 50 [49].

Since maximum stroke volume is largely determined at birth, and does not increase to any significant degree after full growth, the fall in heart rate that occurs with aging directly impacts maximum cardiac output with aging. There can be some compensation by an increase in stroke volume because there usually are some stroke volume reserves but these are limited. In the modeling study, decreasing maximum heart rate to 160 b/min only lowered cardiac output by 2%, but lowering it to 140 b/min decreased maximum cardiac output by 16% with little further change until it was lowered to below 120 b/min.

Regulation of blood pressure during exercise turned out to be very complicated. A key variable for the increase in cardiac output was the increase in fractional flow to the muscle bed. This could occur by just decreasing the arterial resistance going to the muscle vasculature, by increasing the resistance to the non-muscle splanchnic vasculature, or by some combination of both. If the fractional flow to the muscle only was produced by decreasing the arterial resistance in that region, and without muscle contractions, arterial pressure markedly fell in the model because of the increased trapped blood in muscle veins. When muscle contractions were added, the normal volume distribution was restored and blood pressure markedly increased above expected levels. It was then necessary to decrease splanchnic resistance in the model to obtain the expected arterial pressure. Higher arterial pressures also increase the fractional flow to the muscle compartment because the time constant of inflow is much faster to muscle because of its lower arterial resistance. It is the ratio of the arterial resistances between the two compartments that ultimately counts for the slope of venous return, so we found that we had to adjust the two arterial resistances to obtain known values. Failure of compensation of the arterial resistances to the splanchnic and muscle compartments could lead to clinically significant exercise induced hypertension or even hypotension in some cases.

Additional factors: There were no atria in the modeling study. In unpublished studies we have added atrium. Under resting conditions, atrial contractions had little effect on cardiac output but with all the processes in place at peak performance, atrial contractions added as much as a 10–20% further increase in cardiac output.

Besides the muscle pump, there are two other pumping mechanisms that can increase venous return and cardiac output during exercise. The descent of the diaphragm during the vigorous respiratory efforts at peak performance can transiently raise abdominal pressure [50, 51] which will compress the splanchnic venous compartment and increase the rate of venous return from this region. In the thorax, the inspiratory fall in pleural pressure can increase venous return although this effect only can work by lowering venous pressure to atmospheric pressure. Below that the Starling resistor mechanism limits any further increase in the rate of return. Active expiration can increase intrathoracic pressure, which could aid left ventricular ejection, but this benefit is more likely offset by the positive pleural pressure decreasing venous return. The final effect would depend upon the balance of potential recruitment of volume from the splanchnic compartment versus the inhibition to flow in the thorax.

5. Conclusion

Maximum cardiac output is a key determinant of maximum aerobic performance [4]. A fundamental principle in the circular circulatory system is that what goes out per time must come back at the same rate. Thus, the maximum ejected stroke volume per beat must be matched by an equal stroke return. The determinants of stroke return often are under-appreciated in discussions of the limits of aerobic performance. I have reviewed these factors in detail based on an analysis obtained from a computational model of the mechanics of the circulation at maximal exercise [41]. The role of the muscle pump was very evident in the analysis. Besides the obvious importance of muscles for performing work, muscle contractions play an essential role in increasing cardiac output by decompressing the volume that otherwise would accumulate in the muscle venous compartment, and by speeding

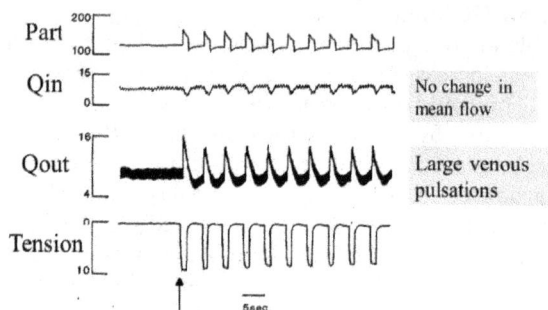

Figure 3.
Effect of muscle contractions on blood flow in isolated gastrocnemius. The gastrocnemius muscle of a dog was isolated, maximally dilated and perfused with a constant flow pump. The top line is arterial pressure (Part, mmHg), the second is inflow (Qin, ml/min measured with an electromagnetic flow probe), the third is outflow (Qout, ml/min) and the fourth the generated longitudinal muscle tension (T, gm). Muscle contractions raised arterial pressure. There was a small transient fall in Qin because the contractions obstruct flow. There are large increases in Qout with the contraction, which make it look like more total flow went out but Qin is constant so the flow is unchanged. This tracing is the same as that of Folkow et al. [52] who based on the venous pulsations concluded that muscle contractions increase muscle blood flow but they did not measure Qin which indicates that mean flow does not change. Reproduced with permission from Naamani et al. Eur J Appl Physiol 1995 [44].

up the return of blood to heart. These roles of contracting muscle in determining cardiac output and decompressing the muscle vascular beds could be a fruitful area for further investigations with implications for maximizing athletic performance, as well rehabilitating persons with reduced muscle capacity. Issues that should be important are ideal rates of contraction and force of contraction needed to have training effect. In our modeling study, increasing contraction rates from 50 to 150 contractions/minute or force of contraction of greater than 0.25 mm Hg/ml did not change cardiac output, presumably because the veins were maximally compressed, but this was an idealized assessment and treated all muscles as one. The rate and force factors could be a much more important factor in arm versus leg exercise, and in debilitated and elderly patients with limited capacity for production of muscle force.

With normal cardiac function, the heart can handle what comes back as evident by maintenance of a low Pra. However, the heart does not have volume reserves that it can use to increase rate of return to the heart by increasing MSFP. Increased cardiac output thus is very dependent upon the increased of the return function and the effect of muscle contractions. On the other side, muscle performance is very dependent upon the delivered blood flow so that flow from the muscle and into it are intricately connected for optimal performance (**Figure 3**).

Author details

Sheldon Magder[1,2]

1 Department of Critical Care, McGill University, Montreal, Quebec, Canada

2 Department of Physiology, McGill University, Montreal, Quebec, Canada

*Address all correspondence to: sheldon.magder@mcgill.ca

IntechOpen

References

[1] Mitchell JH, Sproule BJ, Chapman CB. The physiological meaning of the maximal oxygen intake test. The Journal of Clinical Investigation. 1958;**37**(4):538-547

[2] Åstrand P-O, Cuddy TE, Saltin B, Stenberg J. Cardiac output during submaximal and maximal work. Journal of Applied Physiology. 1964;**19**(2): 268-274

[3] Ekblom B, Hermansen L. Cardiac output in athletes. Journal of Applied Physiology. 1968;**25**(5):619-625

[4] Saltin B, Calbet JA. Point: In health and in a normoxic environment, VO2 max is limited primarily by cardiac output and locomotor muscle blood flow. Journal of Applied Physiology. 2006;**100**(2):744-745

[5] Hsia CC. Respiratory function of hemoglobin. The New England Journal of Medicine. 1998;**338**(4):239-247

[6] Astrand PO, Rodahl K. Physiological bases of exercise. Textbook of Work Physiology. Montreal: McGraw-Hill; 1977

[7] Laughlin MH, Armstrong RB. Muscle blood flow during locomotory exercise. Exercise Sport Science Review. 1985;**13**:95-136

[8] Armstrong RB, Laughlin MH. Blood flows within and among rat muscles as a function of time during high speed treadmill exercise. Journal of Physiology. 1983;**344**:189-208

[9] Clausen JP. Circulatory adjustments to dynamic exercise and effect of physical training in normal subjects and in patients with coronary artery disease. Progress in Cardiovascular Diseases. 1976;**18**(6):459-495

[10] Saltin B, Blomqvist G, Mitchell JH, Johnson RL Jr, Wildenthal K,

Chapman CB. Response to exercise after bed rest and after training. Circulation. 1968;**38**(5 Suppl):Vii1-Vi78

[11] Wang E, Cruz C, Pettersen MR, Hoff J, Helgerud J. Comparison of thoracic bioimpedance with acetylene uptake for measuring cardiac output. International Journal of Sports Medicine. 2014;**35**(11):925-932

[12] Kjellberg SR, Rudhe U, Sjostrand T. The correlation of the cardiac volume to the surface area of the body, the blood volume and the physical capacity for work. Cardiologia. 1949;**14**(6):371-373

[13] McGuire DK, Levine BD, Williamson JW, Snell PG, Blomqvist CG, Saltin B, et al. A 30-year follow-up of the Dallas Bedrest and Training Study: II. Effect of age on cardiovascular adaptation to exercise training. Circulation. 2001;**104**(12): 1358-1366

[14] Levy MN. The cardiac and vascular factors that determine systemic blood flow. Circulation Research. 1979;**44**(6): 739-747

[15] Guyton AC. Determination of cardiac output by equating venous return curves with cardiac response curves. Physiological Reviews. 1955;**35**:123-129

[16] Patterson SW, Starling EH. On the mechanical factors which determine the output of the ventricles. The Journal of Physiology. 1914;**48**(5):357-379

[17] Magder S. Volume and its relationship to cardiac output and venous return. Critical Care. 2016;**20**:271

[18] Magder S. An approach to hemodynamic monitoring: Guyton at the beside. Critical Care. 2012;**16**:236-243

[19] Bishop VS, Stone HL, Guyton AC. Cardiac function curves in conscious

dogs. The American Journal of Physiology. 1964;**207**(3):677-682

[20] Guyton AC, Polizo D, Armstrong GG. Mean circulatory filling pressure measured immediately after cessation of heart pumping. The American Journal of Physiology. 1954;**179**(2):261-267

[21] Lindsey AW, Guyton AC. Continuous recording of pulmonary blood volume: Pulmonary pressure and volume changes. The American Journal of Physiology. 1959;**197**:959-962

[22] Guyton AC, Lindsey AW, Kaufman BN. Effect of mean circulatory filling pressure and other peripheral circulatory factors on cardiac output. The American Journal of Physiology. 1955;**180**:463-468

[23] Rothe CF. Reflex control of veins and vascular capacitance. Physiology Review. 1983;**63**(4):1281-1295

[24] Drees J, Rothe C. Reflex venoconstriction and capacity vessel pressure-volume relationships in dogs. Circulation Research. 1974;**34**:360-373

[25] Magder S, De Varennes B. Clinical death and the measurement of stressed vascular volume. Critical Care Medicine. 1998;**26**:1061-1064

[26] Deschamps A, Fournier A, Magder S. Influence of neuropeptide Y on regional vascular capacitance in dogs. The American Journal of Physiology. 1994;**266**:H165-HH70

[27] Deschamps A, Magder S. Effects of heat stress on vascular capacitance. The American Journal of Physiology. 1994;**266**:H2122-H21H9

[28] Deschamps A, Magder S. Baroreflex control of regional capacitance and blood flow distribution with or without alpha adrenergic blockade. Journal of Applied Physiology. 1992;**263**: H1755-H1H63

[29] Permutt S, Riley S. Hemodynamics of collapsible vessels with tone: The vascular waterfall. Journal of Applied Physiology. 1963;**18**(5):924-932

[30] Notarius CF, Magder S. Central venous pressure during exercise: Role of muscle pump. Canadian Journal of Physiology and Pharmacology. 1996;**74**(6):647-651

[31] Notarius CF, Levy RD, Tully A, Fitchett D, Magder S. Cardiac vs. non-cardiac limits to exercise following heart transplantation. American Heart Journal. 1998;**135**:339-348

[32] Jensen RL, Yanowitz FG, Crapo RO. Exercise hemodynamics and oxygen delivery measurements using rebreathing techniques in heart transplant patients. The American Journal of Cardiology. 1991;**68**(1):129-133

[33] Cerretelli P, Marconi C, Meyer M, Ferretti G, Grassi B. Gas exchange kinetics in heart transplant recipients. Chest. 1992;**101**(5, Supplement): 199S-205S

[34] Krogh A. The regulation of the supply of blood to the right heart. Skandinavian Arch Physiology. 1912;**27**:227-248

[35] Permutt S, Caldini P. Regulation of cardiac output by the circuit: Venous return. In: Boan J, Noordergraaf A, Raines J, editors. Cardiovascular System Dynamics. Cambridge, MA and London England: MIT Press; 1978. pp. 465-479

[36] Caldini P, Permutt S, Waddell JA, Riley RL. Effect of epinephrine on pressure, flow, and volume relationships in the systemic circulation of dogs. Circulation Research. 1974;**34**:606-623

[37] Green JF. Pressure-flow and volume-flow relationships of the systemic circulation of the dog. The American Journal of Physiology. 1975;**229**(3):761-769

[38] Mitzner W, Goldberg H. Effects of epinephrine on resistive and compliant properties of the canine vasculature. Journal of Applied Physiology. 1975;**39**(2):272-280

[39] Sagawa K. The ventricular pressure-volume diagram revisited. Circulation Research. 1978;**43**:677-687

[40] Green JF. Mechanism of action of isoproterenol on venous return. The American Journal of Physiology. 1977;**232**(2):H152-H1H6

[41] Magder S, Famulari G, Gariepy B. Periodicity, time constants of drainage, and the mechanical determinants of peak cardiac output during exercise. Journal of Applied Physiology. 2019;**127**(6):1611-1619

[42] Astrand PO. Quantification of exercise capability and evaluation of physical capacity in man. In: Sonnenblick EH, Lesch M, editors. Exercise and Heart Disease. A progress in Cardiovascular Desease Reprint. New York: Grune and Stratton; 1976. pp. 87-103

[43] Magder S. Venous mechanics of contracting gastrocnemius muscle and the muscle pump theory. Journal of Applied Physiology. 1995;**79**(6): 1930-1935

[44] Naamani R, Hussain SNA, Magder S. The mechanical effects of contractions on muscle blood flow. European Journal of Applied Physiology. 1995;**71**:102-112

[45] Folkow B, Haglund U, Jodal M, Lundgren O. Blood flow in the calf muscle of man during heavy rhythmic exercise. Acta Physiologica Scandinavica. 1971;**81**:157-163

[46] Magder S. Starling resistor versus compliance. Which explains the zero-flow pressure of a dynamic arterial pressure-flow relation? Circulation Research. 1990;**67**:209-220

[47] McGuire DK, Levine BD, Williamson JW, Snell PG, Blomqvist CG, Saltin B, et al. A 30-year follow-up of the Dallas Bedrest and Training Study: I. Effect of age on the cardiovascular response to exercise. Circulation. 2001;**104**(12):1350-1357

[48] Messonnier LA, Emhoff C-AW, Fattor JA, Horning MA, Carlson TJ, Brooks GA. Lactate kinetics at the lactate threshold in trained and untrained men. Journal of Applied Physiology. 2013;**114**(11):1593-1602

[49] Herridge MS, Tansey CM, Matté A, Tomlinson G, Diaz-Granados N, Cooper A, et al. Functional disability 5 years after acute respiratory distress syndrome. The New England Journal of Medicine. 2011;**364**(14):1293-1304

[50] Miller JD, Pegelow DF, Jacques AJ, Dempsey JA. Skeletal muscle pump versus respiratory muscle pump: Modulation of venous return from the locomotor limb in humans. Journal of Physiology. 2005;**563**(Pt 3):925-943

[51] Aliverti A, Bovio D, Fullin I, Dellacà RL, Lo Mauro A, Pedotti A, et al. The abdominal circulatory pump. PLoS One. 2009;**4**(5):e5550

[52] Folkow B, Gaskell P, Waaler BA. Blood flow through limb muscles during heavy rhythmic exercise. Acta Physiologica Scandinavica. 1970;**80**:61-72

Physical Exercise and Immune Response

Chapter 11

Exercise Training in the Spectrum of Breast Cancer

Ana Cristina Corrêa Figueira, Ana Pereira, Luis Leitão,
Rita Ferreira and José Alberto Duarte

Abstract

Exercise training and regular physical activity have been mentioned as one of the non-pharmacological approaches to enhance breast cancer outcomes. Such evidence encourages health professionals to recommend it as an adjuvant in treatment conditions to improve cardiorespiratory fitness that, can increase the rate of completion of pharmacologic therapies, reduce cancer-related fatigue, and improve muscle strength and quality of life. Research results have highlighted a positive relationship between exercise and breast tumor outcomes, that seem to be dose dependent (the more activity the more protection) and can be mediated through several biological mechanisms. In this chapter, we intend to summarize the current knowledge about the effects of exercise in the regulation of metabolic and steroid hormones, tumor-related inflammation, and the attenuation of cancer-induced muscle wasting, highlighting the exercise designs that can prompt the best results.

Keywords: exercise training, breast cancer, physical activity

1. Introduction

Cancer remains not only one of the most daunting diseases worldwide, but also a major public health concern. Although extensive research has been conducted on cancer prevention, diagnosis, and treatment, statistics of cancer's effects are disheartening [1].

Among cancer's various types, breast cancer is the second-most common and the most common among women, and it is assumed that one of every eight women will develop this type of cancer at some point in their lives [2].

The risk of developing breast cancer encompasses reproductive factors, hormonal factors, genetic alterations, age, race, sex, and factors related to lifestyle [3, 4].

Among the lifestyle factors that increase the risk of breast cancer, alcohol consumption [5], tobacco use [6], unhealthy diet [7], and reduced levels of daily physical activity or exercise are ranked among the most major [8].

Unlike determined genetic factors, regular physical activity and/or regular exercise training (RPA/REX) are modifiable ones [9].

In general, RPA/REX can contribute to overall health, with undeniable benefits for cardiorespiratory fitness (CRF), muscle strength insulin resistance, immune function, and body mass index (BMI) maintenance which can particularly be extended to the prevention of several diseases, including breast cancer [10, 11].

IntechOpen

Nevertheless, the biological mechanisms underlying the protective effect of RPA/ REX on breast cancer remain poorly understood. Biomarkers proposed to support that association include the modulation of circulating levels of both metabolic hormones (e.g., insulin, and insulin-like growth factors, IGFs) and steroid hormones (e.g., estradiol and progesterone), the reduction of pro-inflammatory and anti-inflammatory factors (e.g., interleukins, IL-6 and IL-10; tumor necrosis factor-α TNF- α; C-reactive protein, CRP; adiponectin and IL-1), immune function (e.g., natural killers cells and leukocytes), and oxidative stress (i.e., reactive oxygen species) [8, 12, 13].

A considerable number of reviews have reported epidemiological evidence that establishes a connection between RPA/REX and cancer prevention by associating the amount of exercise performed with a decreased risk of developing cancer. Although the role of RPA/REX following the diagnosis of cancer has received less attention from researchers, its importance in controlling and reducing the side effects of cancer therapy is evident [14]. This evidence inspires health professionals to recommend RPA/REX as an adjuvant to improve cardiorespiratory fitness that, consecutively, can improve the rate of completion of pharmacologic therapies, reduce cancer-related fatigue, and improve the quality of life (QoL) [15].

To enhance the survival of patients with breast cancer early detection and improved treatments are fundamental [16]. Researchers in the exercise-oncology field have several concerns regarding therapies that best suit specific cases, minimize the number of deaths, and reduce recurrence. Thus, targeting the association between active behaviors and the cellular and molecular mechanisms underlying that association has been the main target in recent years [17].

Literature provides sufficient evidence to suggest that RPA/REX, when performed at moderate to vigorous intensity for at least 30 min/day, is safe and well-tolerated by patients both during and after therapy [18]. The American College of Sports Medicine (ACSM) recommends that patients with breast cancer avoid inactivity. They should be as active as tolerable by their conditions and should follow the guidelines for healthy individuals when possible: 150 min/week of exercise training at moderate intensity or 70 min/week of exercise training at a vigorous intensity, combining endurance and resistance exercises [19]. Furthermore, these are also the recommendations supported by the American Cancer Society (ACS) [18]. Nevertheless, the patient's overall status should always be examined to ensure an individually adjusted amount of activity by defining personal thresholds of activity determined on a symptom-based approach [20].

A considerable number of studies have provided sufficient evidence that supports those recommendations. Almost a decade ago, a systematic review with meta-analysis, that considered 14 randomized controlled trials (RCT) involving 715 patients with breast cancer, concluded that resistance exercise training (RET) and aerobic exercise training (AET) increase self-esteem, body composition, physical fitness, and the rate of chemotherapy completion [21]. Two years later, a prospective study that involved more than 4000 patients, disclosed that being active during and after treatment for breast cancer can reduce mortality among women regardless of age, state of the disease, and the body mass index (BMI) [22], while a short after, another study, involving more than 14,000 women, showed that high levels of cardiorespiratory fitness were strongly associated with fewer deaths [23]. Similar results were found in another six prospective cohort studies involving more than 12,000 breast cancer survivors gathered in a meta-analysis; results showed that physical activity after the diagnosis of breast cancer reduced death by 34% and recurrence by 24%, regardless of BMI, while pre-diagnosis physical activity reduced the risk of mortality only in women with a BMI <25 kg/m^2 [24]. In more recent years, Lahart et al. [25] and Lipsett et al. [26] quantified the effects of exercise training in breast cancer outcomes during therapy with two meta-analyses. Both revealed that patients with breast cancer benefited by engaging in exercise training activities. While Lahart et al. [25] determined that a combination of RET and AET affords significant benefits by reducing fatigue, Lipsett et al. [26] reported an inverse relationship between RPA levels and breast cancer-related deaths and recurrence.

The results observed, most of which have stemmed from epidemiological evidence, have highlighted a positive association between RPA/REX and breast tumor-associated deaths and recurrence which, apparently, seems to be dose-dependent, meaning more activity, more protection. Nevertheless, confusion remains about the specific amount of exercise that can induce the greatest outcomes.

In contrast, to the studies in clinical contexts that have provided extensive evidence showing that RPA/REX promotes patients' survival and reduced recurrence, such linearity in animal studies has not been found. Furthermore, although evidence showing a positive relationship between RPA/REX and the development of mammary tumors [27] exists, the opposite has also been reported by several researchers [28].

The up mentioned uncertainty has determined interest in understanding whether RPA/REX has a considerable role in tumorigenesis-related outcomes by modulating tumor behavior [24]. Researchers in this area have sought to confirm a relationship between RPA/REX and concurrent biological changes that can determine better outcomes. That association encompasses the intensity, type, and duration of exercise bouts, possible pathophysiological pathways, and breast cancer-associated mechanisms within the context of the advantageous effects of exercise [29]. The present chapter presents an attempt to summarize the insights that linked exercise training and cancer-associated mechanisms along the breast cancer continuum.

2. Exercise training and glucose-related factors

The modulation of metabolic hormones, including markers of glucose-insulin homeostasis are reported to relate to exercise. An altered cellular metabolism that favors aerobic glycolysis to support rapid cell proliferation and high-energy turnover, are features from which the tumors are recognized. The increased consumption of glucose by some tumors, such as those in the breast, that result in increased lactate production (i.e., the Warburg effect), is a well-described mechanism [30]. It seems that limiting glucose availability should restrict the capacity of growth factors to preserve cellular viability, thereby leading to cell death, is a process in which RPA/REX might be crucial. Further, RPA/REX has been hypothesized as an inducer of perturbations in the insulin–glucose axis, enhanced insulin sensitivity, and therefore, promoting a reduction in the circulating levels of insulin and glucose [31].

In the past few years, several studies, mostly RCT, have sought to elucidate whether RPA/REX acts in the modulation of glucose-related factors improving the outcomes in women with breast cancer. In an RCT conducted by Fairey et al. [32] that aimed to determine the effects of REX in glucose-related markers of 53 postmenopausal breast cancer survivors, the women that trained on cycle ergometers for 15 weeks (3 days/week for 35 min) at a moderate intensity have not shown significant changes in fasting insulin, glucose, insulin resistance, or Insulin-like growth factor binding protein-1 (IGFBP). However, the exercise training markedly improved the levels of Insulin-like growth factor (IGF) and Insulin-like growth factor binding protein-3 (IGFBP-3). Theoretically, increases in IGF-1 imply improvements in cell division and the inhibition of cell death [33]; however, also theoretically, because IGFBP-3 is responsible for binding the majority of IGF-1, increased levels of IGFBP-3 should be a good sign. In a different exercise paradigm, Schmitz et al. [34] also led an RCT with 85 postmenopausal breast cancer survivors who experienced a twice-weekly (60 min/session) weight training for one year. The training sessions were supervised for 6 months, and all the participants learned how to work alone and how to improve their workload. Thereafter, and for another 6 months, they continued the training unsupervised. Positive results were found regarding IGF-2 levels, but no evidence was detected concerning improvements in insulin sensitivity and glucose levels [34]. The above-mentioned results, which are both discouraging and challenging, do not indicate that REX can effectively improve glucose levels, insulin levels, or insulin resistance. Although a positive relationship regarding IGF-2 levels was found, this can be related to the different exercise designs presented in both studies, namely the different lengths of exercise exposure, the differences in the types of exercise, or the reductions in women's BMI, that was reported in the latter but not in the former study.

A few years later, 101 sedentary and overweight breast cancer survivors, were randomly assigned by Ligibel et al. [35] to either a 16-week program of unsupervised AET (90 min/week) combined with RET (2 days/week for 50 min with supervision) or to a control group. Aiming to analyze the influence of REX on insulin concentrations, positive changes were reported for fasting insulin with some evidence for improvement in insulin resistance but not for fasting glucose [35]. Likewise, Irwin et al. [31] studied 75 postmenopausal breast cancer survivors who were subjected to an exercise program involving three weekly supervised sessions and twice-weekly unsupervised sessions (30 min of moderate AET for 6 months). Women with higher exercise levels were reported to show a decrease in insulin, IGF-1, and IGFBP-3. Interestingly and despite other evidence [36], the authors assumed that the decrease in IGFBP-3 was probably related to a similar reduction of IGF-1 levels.

In the same line of research, an RCT conducted by Guinan et al. [37] with 26 breast cancer survivors, that combined supervised exercise with a home-based program, performed twice a week for 60 min during 8 weeks, did not find any evidences of positive changes in the circulating levels of glucose and insulin. Likewise, Thomas et al. [38] included 65 postmenopausal breast cancer survivors in an exercise program that combined supervised (3 days/week) and unsupervised (2 days/week) 30-min training sessions of moderate exercise, aimed to achieve 150 min/week during a period of 6-month. Significant results were found, translated into a reduction in fasting glucose among the more active women (>120 min/week). Again, the different results are probably related to the different exercise designs used, which prevent us to reach a satisfactory conclusion about the type and amount of exercise to prescribe for breast cancer patients and that best contribute to positive outcomes. Two recently published meta-analyses did not highlight any clues to clarify this matter. Albeit demonstrating that exercise can reduce fasting insulin levels and IGFs [39] in breast cancer survivors, differences in the exercise training protocols prevent the attempt of a strength subgroup analysis in one of them [39]. On the other, the heterogeneity in exercise designs was mentioned as a limitation, and no subgroup analysis was performed [40]. Comparably, to trends in human studies, contrasting results also characterize preclinical data. Several reports have associated RPA/REX with increases in the levels of glucose-related factors [41], whereas others have related the opposite [42]. Additionally, and similarly to the research in human populations, the use of different exercise training programs prevents any clear understanding of the amount of exercise needed to enhance the glucose-related indicators.

3. Exercise training and inflammation-related factors

Another key factor in the improvement and progression of breast cancer is chronic inflammation. RPA/REX could neutralize the permanent state of inflammation by promoting a systemic anti-inflammatory environment. One of the mechanisms by which exercise can reduce cancer-related inflammation is through the increasing levels of anti-inflammatory myokines produced by the working skeletal muscles. Such decreases in cancer-related inflammation could be related to the frequency, intensity, type, and duration, of the exercise training sessions. In fact, it seems that the intensity may be the key. Higher levels of RPA/REX intensity (i.e., moderate or vigorous) can promote the reduction of the circulating levels of proinflammatory cytokines and improve immune function [43].

The collected data relating to cancer-induced inflammation and exercise training in human patients can illustrate the urgency for more studies.

In an RCT with 52 breast cancer survivors, Fairey et al. [44] exposed the participants to 15 weeks of moderate exercise in cycle ergometer and observed significant enhancements in immune function expressed by exercise-induced natural killer cell activity, but any association between REX and anti-inflammatory interleukins (i.e., IL-4 and IL-10), pro-inflammatory interleukins (i.e., IL-1 and IL-6), tumor necrosis factor-α (TNF-α) and cytokines, were not detected. With the same intervention group, the authors reported positive associations between REX and the C reactive protein (CRP) levels. Equally, Hutnick et al. [45] in a study involving 28 breast cancer survivors exposed to moderate treadmill exercise program combined with resistance training for 6 months, did not observe any association between REX and plasma IL-6 levels and interferon-gamma (IFN-γ). Nevertheless, the improved activation of lymphocytes in women who exercised showcased a positive relationship between immune function and exercise. In another RCT, conducted by Gómez et al. [46] with 16 breast cancer survivors exposed to an 8-week, a three-times-weekly program that combined resistance exercise training with aerobic training, did not detect significant changes in their inflammation-related systemic markers (e.g., IL-6, IL-10, and TNF-α). Rogers et al. [47] after a 3-month training program combining AET and RET in breast cancer survivors, found a positive relationship between leptin levels and REX, showing relevant evidence of the benefits induced by REX in proinflammatory cytokines (i.e., IL-6 and TNF-α).

Likewise, Campbell et al. [48] did not find significant associations between a 24-week home-based program of moderate exercise and inflammatory markers in 37 postmenopausal breast cancer survivors.

Although the amount of research performed in the last decade, the importance of RPA/REX to induced changes in the systemic repercussions of breast cancer, remains doubtful. Once again, differences among studies can confound the true benefits of exercise training in inflammation-related markers and enhancing immune response.

Trying to overtake this uncertainty, a recent meta-analysis highlighted the positive effects of chronic exercise training in low-grade inflammation in women with breast cancer. The benefits associated with the intervention program duration (>45 minutes/session) and length (>11 weeks) showed that a significant decrease in TNF-α levels were associated with decreased levels of adiposity [49]. Yet encouraging, such results should be interpreted with caution given the number of correlations performed, which in some cases were quite a few.

Similarly, the preclinical data about the advantages of RPA/REX to modulate inflammation are also diverse, including the improved expression of several inflammation biomarkers (e.g., CRP, TNF-α, IL-6, INF-γ, monocyte chemoattractant protein 1 [MCP1], serum amyloid P [SAP], leptin and spleen weight) [50], although other data has reported the opposite, primarily regarding in IL-6 regulation [51].

In that case, the considerable variety of exercise designs and breast tumor models might have significantly influenced the outcomes.

4. Exercise training and estrogen levels along the breast cancer continuum

RPA/REX has been associated with lower levels of circulating estrogen, which could describe its positive association with breast cancer. Cell proliferation and the inhibition of apoptosis via ER (estrogen receptor)-mediated mechanisms have been associated with the circulating levels of estrogen [52]. Apparently, physically active women with breast cancer, shows lower estrogen levels that can improve survival, particularly with tumors overexpressing positive estrogen receptors (ER+) and positive progesterone receptors (PR+), though the lack of data from human studies to support that hypothesis [24].

Using 1970 women from a previous cohort study to assess the levels of RPA/REX self-reported on a questionnaire, Sternfeld et al. [53] found no differences in levels of RPA/REX, tumors' hormonal status, or the number of breast cancer deaths. Curiously, the number of all-cause deaths was markedly lower among women who presented ER+ and PR+ tumors and had engaged in RPA/REX programs of moderate intensity. Irwin et al. [54] measured the self-reported data of 2910 women with breast cancer regarding their RPA/REX behaviors, to examine whether such behaviors influenced mortality, but did not find significant results between the RPA/REX levels and mortality in women with ER+ and negative Human Epithelial Growth Factor Receptor (HER−) tumors. The methodological approach of the study conducted by Chen et al. [55] was identical, but the results differed starkly. The authors observed significant effects between RPA/REX levels and the reduced number of total mortalities only among women with ER− or PR− tumors, but not with ER+ or PR+ ones. Human studies have divergent results and evidence even though most have used the same or similar methodological processes. Of note are the reduced number of studies concerning the relationship between RPA/REX condition and the circulating levels of estrogen in the progression and development of breast cancer, especially because those concentrations are associated with the growth of tumor cells [56].

The lack of data in preclinical research also restrains any understanding of the role of RPA/REX and the circulating levels of progesterone and estrogen. In this field, some studies have shown a positive association [41, 57], between RPA/REX and the circulating levels of sex hormone, while the contrary can be also found [58]. These findings suggest that methodological limitations, including heterogeneous cohorts, small sample sizes, and randomized characteristics, could translate the conflicting results regarding the effect of RPA/REX on breast cancer-associated markers.

A recent systematic review with meta-analysis in pre-clinical data reported considerable improvements in sex hormone concentrations, cancer-induced inflammation, and glucose-related factors among the animals exposed to exercise [59]. Performing vigorous exercise for 85 min per week improved sex hormone levels and reduced systemic inflammation.

5. Exercise training and muscle mass in the breast cancer continuum

The loss of skeletal muscle is a well-documented process in cancer that affects most patients, even though different degrees [60]. The musculoskeletal system provides the basic functions of strength generation, locomotion, and respiration. The protection of muscle mass whether in disease or with health is crucial [61]. The preservation of muscle fiber size and muscle mass depends on protein turnover, in which the balance between protein synthesis and protein breakdown should be maintained. A complex system of signaling pathways regulates that stability, and under pathological conditions, such regulation can be compromised and result in muscle decrease and atrophy [62]. Muscle function in oncology relies on the study of cancer-associated cachexia. Systemic metabolic disorder and inflammation caused by tumors seem to affect protein turnover by promoting wasting in muscle mass involving diminished muscle fiber diameter, reduced protein content, decreased fatigue resistance, and force production [63].

In disease the loss of muscle mass reduces the patients' functional capacity, thus considerable efforts have been made in the past few years to find an effective anticachectic gent, and exercise has been suggested as a possible measure to mitigate or reverse muscle dysfunction and wasting or both [64]. Studies in humans with different types of cancer have reached different conclusions although, most of them, normally favor exercise. Unfortunately, in patients with breast cancer, results are insufficient due to the lower incidence of breast cancer patients who suffer from cachexia. As mentioned in a study conducted by Schmitz et al. [34], with 85 cancer survivors divided into two exercise training groups (immediate treatment and delayed treatment), that performed a twice-weekly (60 m/session) during 12 or 6 months, both groups exhibited increases in lean mass and the results were more expressed in the immediate treatment group, whereas the delayed treatment group did not show differences in lean mass across the study period [34]. Courneya et al. [65] obtained similar results after randomizing into three groups 242 patients with breast cancer receiving ongoing treatments. An exercise group that performed AET with different ergometers of moderate-to-vigorous intensity 45 min/day three times weekly, another exercise group that performed RET involving two sets of 8/12 repetitions in different muscle groups three times weekly, and a control group. Both exercise training groups showed significant results relating to the increased lean mass, after 17 weeks, more expressed in women who received RET treatment [66].

Animal studies have also highlighted the benefits of exercise against the depletion of muscle mass in the cancer context. Al-Majid et al. [67] described the benefits of REX program in the lower limb muscles of tumor-bearing animals subjected to eight sessions of electrical stimulation modeling RET. Puppa et al. [55] also reported the beneficial effects of moderate-intensity REX in the treadmill for 55 min/day 6 days/week for 11 weeks on muscle mass, in cancer-induced cachectic animals overexpressing systemic IL-6.

The study of Franjacomo et al. [68] aimed to examine if the studies involving cachexia could use the model of mammary neoplasms. They determined that the model, involving Ehrlich carcinoma cells inoculation could feature systemic inflammation and the muscle wasting of cachexia in a less aggressive manner suitable for studying new pharmacological approaches. Nevertheless, using an inoculation model of Walker-256 cancer cells subjected rats to a 6-week RET program, Padilha et al. [69] concluded that RET performed prior to tumor implantation prevented the development of cachexia by attenuating tumor-induced systemic proinflammatory conditions, oxidative stress, and damage in the muscles, which suggest the advantages of exercise training prior to tumor onset.

Above the heterogeneity of results and research designs in animal models and clinical conditions, it seems that the severity and incidence of cachexia depend on a reasonable range of factors and can vary according to the site and mass of the tumor, tumor type, interindividual differences in susceptibility to cachexia, and abnormal metabolism or reductions in food intake [70].

Evidence from animal studies demonstrates that RPA/REX might play an important role in attenuating cancer-induced muscle wasting by regulating or inhibiting, if not both, several factors at a molecular level. Apparently, exercise activates a network of transcription factors, kinases, and coregulator proteins that cap change in gene expression and prompt increases in mitochondrial biogenesis, which in turn cause metabolic reprogramming in skeletal muscle. Consistent evidence shows that endurance training induces mitochondrial biogenesis and a fast-to-slow fiber-type switch in skeletal muscles, expressed in type 1 and type 2A fibers [71].

Clearly, additional studies are needed in the context of cancer-induced muscle wasting, to date, no medical intervention has completely reversed cachexia, and no approved drug therapies are available [72]. Nevertheless, according to recent data [73], molecular mechanisms underlying such beneficial effect of exercise seems to be by the contribution of TNF-like weak inducer of apoptosis (TWEAK) signaling to cancer-induced skeletal muscle wasting. The authors concluded that exercise training prevented tumor-induced TWEAK/NF-jB signaling in skeletal muscle with a beneficial impact on fiber cross-sectional area and metabolism. Indeed, 35 weeks of exercise training promoted the upregulation of oxidative complexes. An active lifestyle for the prevention of muscle wasting secondary to breast cancer, highlighting TWEAK/NF-jB signaling as a potential therapeutic target for the preservation of muscle mass.

6. Conclusion

The diversity of designs and protocols of exercise training used in the reviewed documents, create serious difficulties to achieve a clear understanding of the best exercise training designs that better suit greater outcomes for breast cancer patients.

More powerful studies involving similar protocols could enhance the knowledge of the ideal amount of exercise needed in clinical contexts. However, despite the diversity of results reported, evidence of the benefits of exercising after a breast cancer diagnosis does exist.

Exposure to programs of regular exercise training combining AET and RET seems to promote better results along with the moderate-to-vigorous intensity through which the exercise is performed.

Exercise programs to breast cancer patients should include organized and supervised activities, considering a symptom-based approach, tailored to each patient.

Performing vigorous exercise 85 min/week can reduce the levels of systemic inflammation and improve circulating in sex hormone levels in animals, but in human populations, such evidence needs to be supported with more studies. The present chapter has some limitations mainly related to the studies included. The difference in study designs along with the lack of published information in some of them creates a difficult understanding and a more accurate analysis. Future studies should be performed with the knowledge already achieved in mind reporting, when published, the necessary information that allows their replication.

Acknowledgements

The authors gratefully acknowledge the financial support from the Portuguese Foundation for Science and Technology, I.P., Grant/Award Number UIDP/04748/2020 and UIB/00617/2020, from the Polytechnic Institute of Setúbal and from the School of Education.

Author details

Ana Cristina Corrêa Figueira[1,2*], Ana Pereira[1,3], Luis Leitão[1,3], Rita Ferreira[2,4] and José Alberto Duarte[2]

1 Department of Sciences and Technologies/Sport Sciences, Polytechnic Institute of Setúbal, School of Education, Setúbal, Portugal

2 CIAFEL, Research Center in Physical Activity, Exercise, Leisure and Health, Faculty of Sport, University of Porto, Porto, Portugal

3 Life Quality Research Center, Rio Maior, Portugal

4 QOPNA, Department of Chemistry, University of Aveiro, Aveiro, Portugal

*Address all correspondence to: ana.figueira@ese.ips.pt

IntechOpen

References

[1] Siegel RL, Miller KD, Jemal A. Cancer statistics, 2017. CA: A Cancer Journal for Clinicians. 2017;**67**:7-30

[2] Kolak A, Kaminska M, Sygit K, Budny A, Surdyka D, Kukielka-Budny B, et al. Primary and secondary prevention of breast cancer. Annals of Agricultural and Environmental Medicine. 2017;**24**:549-553

[3] Singletary SE. Rating the risk factors for breast cancer. Annals of Surgery. 2003;**237**:474-482

[4] Weinberg R. The Biology of Cancer. Second Edition. New York: Taylor & Francis Group; 2014

[5] Jayasekara H, MacInnis RJ, Room R, English DR. Long-term alcohol consumption and breast, upper aero-digestive tract and colorectal Cancer risk: A systematic review and Meta-analysis. Alcohol and Alcoholism. 2016;**51**:315-330

[6] Catsburg C, Miller AB, Rohan TE. Active cigarette smoking and risk of breast cancer. International Journal of Cancer. 2015;**136**:2204-2209

[7] Michels KB, Mohllajee AP, Roset-Bahmanyar E, Beehler GP, Moysich KB. Diet and breast cancer: A review of the prospective observational studies. Cancer. 2007;**109**:2712-2749

[8] Friedenreich CM, Cust AE. Physical activity and breast cancer risk: Impact of timing, type and dose of activity and population subgroup effects. British Journal of Sports Medicine. 2008;**42**:636-647

[9] Ratnasinghe LD, Modali RV, Seddon MB, Lehman TA. Physical activity and reduced breast cancer risk: A multinational study. Nutrition and Cancer. 2010;**62**:425-435

[10] Friedenreich CM, Neilson HK, Lynch BM. State of the epidemiological evidence on physical activity and cancer prevention. European Journal of Cancer. 2010;**46**:2593-2604

[11] Schmitz KH, Holtzman J, Courneya KS, Masse LC, Duval S, Kane R. Controlled physical activity trials in cancer survivors: A systematic review and meta-analysis. Cancer Epidemiology, Biomarkers & Prevention. 2005;**14**:1588-1595

[12] Campbell KL, McTiernan A. Exercise and biomarkers for cancer prevention studies. The Journal of Nutrition. 2007;**137**:161S-169S

[13] McTiernan A. Mechanisms linking physical activity with cancer. Nature Reviews. Cancer. 2008;**8**:205-211

[14] Courneya KS, McKenzie DC, Mackey JR, Gelmon K, Friedenreich CM, Yasui Y, et al. Effects of exercise dose and type during breast cancer chemotherapy: Multicenter randomized trial. Journal of the National Cancer Institute. 2013;**105**:1821-1192

[15] Sweegers MG, Altenburg TM, Chinapaw MJ, Kalter J, Verdonck-de Leeuw IM, Courneya KS, et al. Which exercise prescriptions improve quality of life and physical function in patients with cancer during and following treatment? A systematic review and meta-analysis of randomized controlled trials. British Journal of Sports Medicine. 2017;**52**:(8)

[16] Sorlie T. Molecular portraits of breast cancer: Tumor subtypes as distinct disease entities. European Journal of Cancer. 2004;**40**:2667-2675

[17] Lynch BM, Neilson HK, Friedenreich CM. Physical activity and

breast cancer prevention. Recent Results in Cancer Research. 2011;**186**:13-42

[18] Rock CL, Doyle C, Demark-Wahnefried W, Meyerhardt J, Courneya KS, Schwartz AL, et al. Nutrition and physical activity guidelines for cancer survivors. CA: a Cancer Journal for Clinicians. 2012;**62**:243-274

[19] Schmitz KH, Courneya KS, Matthews C, Demark-Wahnefried W, Galvao DA, Pinto BM, et al. American College of Sports Medicine roundtable on exercise guidelines for cancer survivors. Medicine and Science in Sports and Exercise. 2010;**42**: 309-1426

[20] Newton RU, Galvao DA. Exercise in prevention and management of cancer. Current Treatment Options in Oncology. 2008;**9**:135-146

[21] McNeely ML, Campbell KL, Rowe BH, Klassen TP, Mackey JR, Courneya KS. Effects of exercise on breast cancer patients and survivors: A systematic review and meta-analysis. CMAJ. 2006;**33**:34-41

[22] Holick CN, Newcomb PA, Trentham-Dietz A, Titus-Ernstoff L, Bersch AJ, Stampfer MJ, et al. Physical activity and survival after diagnosis of invasive breast cancer. Cancer Epidemiology, Biomarkers & Prevention. 2008;**17**:379-386

[23] Peel JB, Sui X, Adams SA, Hebert JR, Hardin JW, Blair SN. A prospective study of cardiorespiratory fitness and breast cancer mortality. Medicine and Science in Sports and Exercise. 2009;**41**:742-748

[24] Ibrahim EM, Al-Homaidh A. Physical activity and survival after breast cancer diagnosis: meta-analysis of published studies. Medical Oncology. 2011;**28**:753-765

[25] Lahart IM, Metsios GS, Nevill AM, Carmichael AR. Physical activity, risk of death and recurrence in breast cancer survivors: A systematic review and meta-analysis of epidemiological studies. Acta Oncologica. 2015;**54**: 635-654

[26] Lipsett A, Barrett S, Haruna F, Mustian K, O'Donovan A. The impact of exercise during adjuvant radiotherapy for breast cancer on fatigue and quality of life: A systematic review and meta-analysis. Breast. 2017;**32**:144-155

[27] Westerlind KC, McCarty HL, Schultheiss PC, Story R, Reed AH, Baier ML, et al. Moderate exercise training slows mammary tumour growth in adolescent rats. European Journal of Cancer Prevention. 2003;**12**:281-287

[28] Malicka I, Siewierska K, Pula B, Kobierzycki C, Haus D, Paslawska U, et al. The effect of physical training on the N-methyl-N-nitrosourea-induced mammary carcinogenesis of Sprague-Dawley rats. Experimental Biology and Medicine. 2015;**240**:308-1415

[29] Courneya KS, Friedenreich CM. Physical activity and cancer: An introduction. Recent Results in Cancer Research. 2011;**186**:1-10

[30] Warburg O. On the origin of cancer cells. Science. 1956;**123**:309-314

[31] Irwin ML, Varma K, Alvarez-Reeves M, Cadmus L, Wiley A, Chung GG, et al. Randomized controlled trial of aerobic exercise on insulin and insulin-like growth factors in breast cancer survivors: The Yale exercise and survivorship study. Cancer Epidemiology, Biomarkers & Prevention. 2009;**18**:306-335

[32] Fairey AS, Courneya KS, Field CJ, Bell GJ, Jones LW, Mackey JR. Effects of exercise training on fasting insulin,

insulin resistance, insulin-like growth factors, and insulin-like growth factor binding proteins in postmenopausal breast cancer survivors: A randomized controlled trial. Cancer Epidemiology, Biomarkers & Prevention. 2003;**12**:721-727

[33] Schernhammer ES, Holly JM, Hunter DJ, Pollak MN, Hankinson SE. Insulin-like growth factor-I, its binding proteins (IGFBP-1 and IGFBP-3), and growth hormone and breast cancer risk in the nurses health study II. Endocrine-Related Cancer. 2006;**13**:583-592

[34] Schmitz KH, Ahmed RL, Hannan PJ, Yee D. Safety and efficacy of weight training in recent breast cancer survivors to alter body composition, insulin, and insulin-like growth factor axis proteins. Cancer Epidemiology, Biomarkers & Prevention. 2005;**14**:1672-1162

[35] Ligibel JA, Campbell N, Partridge A, Chen WY, Salinardi T, Chen H, et al. Impact of a mixed strength and endurance exercise intervention on insulin levels in breast cancer survivors. Journal of Clinical Oncology. 2008;**26**:907-912

[36] Samani AA, Yakar S, LeRoith D, Brodt P. The role of the IGF system in cancer growth and metastasis: Overview and recent insights. Endocrine Reviews. 2007;**28**:20-47

[37] Guinan E, Hussey J, Broderick JM, Lithander FE, O'Donnell D, Kennedy MJ, et al. The effect of aerobic exercise on metabolic and inflammatory markers in breast cancer survivors--a pilot study. Support Care Cancer. 2013;**21**:1983-1992

[38] Thomas GA, Alvarez-Reeves M, Lu L, Yu H, Irwin ML. Effect of exercise on metabolic syndrome variables in breast cancer survivors. International Journal of Endocrinology. 2013;**2013**:168797

[39] Meneses-Echavez JF, Jimenez EG, Rio-Valle JS, Correa-Bautista JE, Izquierdo M, Ramirez-Velez R. The insulin-like growth factor system is modulated by exercise in breast cancer survivors: A systematic review and meta-analysis. BMC Cancer. 2016;**16**:682

[40] Kang DW, Lee J, Suh SH, Ligibel J, Courneya KS, Jeon JY. Effects of exercise on insulin, IGF Axis, Adipocytokines, and inflammatory markers in breast Cancer survivors: A systematic review and Meta-analysis. Cancer Epidemiology, Biomarkers & Prevention. 2017;**26**:355-365

[41] Zhu Z, Jiang W, Zacher JH, Neil ES, McGinley JN, Thompson HJ. Effects of energy restriction and wheel running on mammary carcinogenesis and host systemic factors in a rat model. Cancer Prevention Research. 2012;**5**:414-422

[42] Gillette CA, Zhu Z, Westerlind KC, Melby CL, Wolfe P, Thompson HJ. Energy availability and mammary carcinogenesis: Effects of calorie restriction and exercise. Carcinogenesis. 1997;**18**:119-1188

[43] Pierce BL, Neuhouser ML, Wener MH, Bernstein L, Baumgartner RN, Ballard-Barbash R, et al. Correlates of circulating C-reactive protein and serum amyloid a concentrations in breast cancer survivors. Breast Cancer Research and Treatment. 2009;**32**:155-167

[44] Fairey AS, Courneya KS, Field CJ, Bell GJ, Jones LW, Martin BS, et al. Effect of exercise training on C-reactive protein in postmenopausal breast cancer survivors: A randomized controlled trial. Brain, Behavior, and Immunity. 2005;**19**:381-388

[45] Hutnick NA, Williams NI, Kraemer WJ, Orsega-Smith E, Dixon RH, Bleznak AD, et al. Exercise and lymphocyte activation following chemotherapy for breast cancer.

Medicine and Science in Sports and Exercise. 2005;**37**:1827-1195

[46] Gomez AM, Martinez C, Fiuza-Luces C, Herrero F, Perez M, Madero L, et al. Exercise training and cytokines in breast cancer survivors. International Journal of Sports Medicine. 2011;**32**:461-467

[47] Rogers LQ, Fogleman A, Trammell R, Hopkins-Price P, Vicari S, Rao K, et al. Effects of a physical activity behavior change intervention on inflammation and related health outcomes in breast cancer survivors: Pilot randomized trial. Integrative Cancer Therapies. 2013;**12**:323-335

[48] Campbell KL, Van Patten CL, Neil SE, Kirkham AA, Gotay CC, Gelmon KA, et al. Feasibility of a lifestyle intervention on body weight and serum biomarkers in breast cancer survivors with overweight and obesity. Journal of the Academy of Nutrition and Dietetics. 2012;**20**:559-567

[49] Meneses-Echavez JF, Correa-Bautista JE, Gonzalez-Jimenez E, Schmidt Rio-Valle J, Elkins MR, Lobelo F, et al. The effect of exercise training on mediators of inflammation in breast Cancer survivors: A systematic review with Meta-analysis. Cancer Epidemiology, Biomarkers & Prevention. 2016;**25**:1009-1017

[50] Goh J, Tsai J, Bammler TK, Farin FM, Endicott E, Ladiges WC. Exercise training in transgenic mice is associated with attenuation of early breast cancer growth in a dose-dependent manner. PLoS One. 2013;**8**:e2123

[51] Thompson HJ, Wolfe P, McTiernan A, Jiang W, Zhu Z. Wheel running-induced changes in plasma biomarkers and carcinogenic response in the 1-methyl-1-nitrosourea-induced rat model for breast cancer. Cancer Prevention Research. 2010;**3**:344-1492

[52] Lewis-Wambi JS, Jordan VC. Estrogen regulation of apoptosis: How can one hormone stimulate and inhibit? Breast Cancer Research. 2009;**11**:206

[53] Sternfeld B, Weltzien E, Quesenberry CP Jr, Castillo AL, Kwan M, Slattery ML, et al. Physical activity and risk of recurrence and mortality in breast cancer survivors: Findings from the LACE study. Cancer Epidemiology, Biomarkers & Prevention. 2009;**18**:87-95

[54] Irwin ML, McTiernan A, Manson JE, Thomson CA, Sternfeld B, Stefanick ML, et al. Physical activity and survival in postmenopausal women with breast cancer: Results from the women's health initiative. Cancer Prevention Research (Philadelphia, Pa.). 2011;**4**:522-529

[55] Chen X, Lu W, Zheng W, Gu K, Matthews CE, Chen Z, et al. Exercise after diagnosis of breast cancer in association with survival. Cancer Prevention Research (Philadelphia, Pa.). 2011;**4**:309-1418

[56] McTiernan A, Rajan KB, Tworoger SS, Irwin M, Bernstein L, Baumgartner R, et al. Adiposity and sex hormones in postmenopausal breast cancer survivors. Journal of Clinical Oncology. 2003;**21**:1961-1966

[57] Isanejad A, Alizadeh AM, Amani Shalamzari S, Khodayari H, Khodayari S, Khori V, et al. MicroRNA-206, let-7a and microRNA-21 pathways involved in the anti-angiogenesis effects of the interval exercise training and hormone therapy in breast cancer. Life Sciences. 2016;**151**: 30-40

[58] Faustino-Rocha AI, Gama A, Oliveira PA, Alvarado A, Neuparth MJ, Ferreira R, et al. Effects of lifelong exercise training on mammary tumorigenesis induced by MNU in female Sprague-Dawley rats. Clinical and Experimental Medicine. 2016;**17**:151-160

[59] Figueira A, Cortinhas A, Soares JP, Leitão FC, Ferreira RP, Duarte JA. Exercise-induced changes in systemic biomarkers of breast cancer: Systematic review with meta-analysis. International Journal of Sports Medicine. 2018;**39**(05):327-342

[60] Al-Majid S, Waters H. The biological mechanisms of cancer-related skeletal muscle wasting: The role of progressive resistance exercise. Biological Research for Nursing. 2008;**10**:7-20

[61] Argiles JM, Campos N, Lopez-Pedrosa JM, Rueda R, Rodriguez-Manas L. Skeletal muscle regulates metabolism via Interorgan crosstalk: Roles in health and disease. Journal of the American Medical Directors Association. 2016;**17**(9):789-796

[62] Sandri M. Signaling in muscle atrophy and hypertrophy. Physiology (Bethesda). 2008;**23**:160-170

[63] Allred DC. Issues and updates: Evaluating estrogen receptor-alpha, progesterone receptor, and HER2 in breast cancer. Modern Pathology. 2010;**23**(Suppl 2):S52-S59

[64] Bowen TS, Schuler G, Adams V. Skeletal muscle wasting in cachexia and sarcopenia: Molecular pathophysiology and impact of exercise training. Journal of Cachexia, Sarcopenia and Muscle. 2015;**6**:197-128

[65] Courneya KS, Segal RJ, Mackey JR, Gelmon K, Reid RD, Friedenreich CM, et al. Effects of aerobic and resistance exercise in breast cancer patients receiving adjuvant chemotherapy: A multicenter randomized controlled trial. Journal of Clinical Oncology. 2007;**25**:4396-4404

[66] al-Majid S, McCarthy DO. Resistance exercise training attenuates wasting of the extensor digitorum longus muscle in mice bearing the colon-26 adenocarcinoma. Biological Research for Nursing. 2001;**2**:155-166

[67] Puppa MJ, White JP, Velazquez KT, Baltgalvis KA, Sato S, Baynes JW, et al. The effect of exercise on IL-6-induced cachexia in the Apc (min/+) mouse. Journal of Cachexia, Sarcopenia and Muscle. 2012;**3**:117-137

[68] Frajacomo FT, de Souza Padilha C, Marinello PC, Guarnier FA, Cecchini R, Duarte JA, et al. Solid Ehrlich carcinoma reproduces functional and biological characteristics of cancer cachexia. Life Sciences. 2016;**162**:47-53

[69] Padilha CS, Borges FH, Mendes C, da Silva LE, Frajacomo FTT, Jordao AA, et al. Resistance exercise attenuates skeletal muscle oxidative stress, systemic pro-inflammatory state, and cachexia in Walker-256 tumor-bearing rats. Applied Physiology, Nutrition, and Metabolism. 2017;**42**:916-923

[70] Argiles JM, Busquets S, Lopez-Soriano FJ, Costelli P, Penna F. Are there any benefits of exercise training in cancer cachexia? Journal of Cachexia, Sarcopenia and Muscle. 2012;**3**:73-76

[71] Arany Z, Lebrasseur N, Morris C, Smith E, Yang W, Ma Y, et al. The transcriptional coactivator PGC-1beta drives the formation of oxidative type IIX fibers in skeletal muscle. Cell Metabolism. 2007;**5**:35-46

[72] Baracos VE. Bridging the gap: Are animal models consistent with clinical cancer cachexia? Nature Reviews. Clinical Oncology. 2018;**15**:197-198

[73] Padrão AI, Figueira ACC, Faustino-Rocha AI, Gama A, Loureiro MM, Neuparth MJ, et al. Long-term exercise training prevents mammary tumorigenesis-induced muscle wasting in rats through the regulation of TWEAK signaling. Acta Physiologica (Oxford, England). 2017;**219**:23-813

Chapter 12

Physical Activity and Vaccine Response

Kotaro Suzuki

Abstract

Over the past decade, numerous research studies have shown that the immune system's capacity for creating antibodies after getting vaccinated is better in those who exercise are physically active. Authoritative studies show that exercise is an important ally of the vaccine, amplifying its effectiveness. The immune response to vaccines is usually lower in the elderly population. Several strategies have been used to help overcome this problem. Recently, studies in humans and animals have shown that exercise increases antigen-specific blood antibody levels following vaccination. Exercise has been considered as an effective way to improve vaccine response in the elderly population. In this chapter, we will discuss the effect of exercise on vaccine response. This study summarizes the current understanding of exercise and antibody production. In order to develop intervention strategies, it will be necessary to further elucidate the predisposing factors and mechanisms behind exercise induce antibody response.

Keywords: exercise, physical activity, vaccine response, antibody

1. Introduction

In our daily life, there are harmful viruses, bacteria, and other microorganisms that can invade the human body and cause illness. However, the human body has a mechanism to prevent pathogens from causing disease once they have invaded the body. This mechanism is called "immunity." Immunity is a powerful defense mechanism that protects us from disease by recognizing pathogens in the body and killing them. Vaccines make use of this mechanism.

Vaccination is one of the most successful public health interventions in preventing infectious diseases and reducing the mortality and morbidity associated with these diseases. The main aim of vaccination is to prevent pathogen-specific infections. The result is to prevent people from becoming seriously ill and dying. On the other hand, aging is the biggest risk factor for impaired immunological health and reduced vaccine efficacy. To enhance the vaccine response, the vaccine itself needs to be modified or behavioral interventions need to be found that alter host factors to enhance the vaccine response.

Exercise improves antibody response to vaccine in human study [1]. Despite the potential beneficial role of exercise on immune responses to vaccination, the underlying mechanisms remain understudied. Based on the studies above, it is generally accepted that prolonged intense exercise is detrimental [2], whilst continuous moderate-intensity exercise is beneficial to immune function [3]. Exercise is a cost-effective behavioral intervention to enhance immune function. Exercise may have a beneficial effect on the immune response to vaccination in elderly population. The two main questions of interest are: (1) how does exercise benefit the effectiveness of vaccination; and (2) what kind of exercise, for how long and at what intensity, would be beneficial in elderly population?

The primary goal of the review presented in this chapter is to provide a better understanding of exercise and vaccine response. This chapter is divided into three parts, with the first section summarizing basic knowledge about vaccines and antibody production. In the second part, we focus on the latest insights into the mechanism of exercise-induced increase in antibody concentration. This section describes the intensity of exercise, the duration of exercise, endogenous opioids, IgG half-life that are modulated by exercise. The third part presents information on our current understanding on immune senescence and effects of exercise on vaccine response in older adult.

2. Vaccination

2.1 What is in a vaccine?

Vaccination is regarded one of the greatest medical discoveries of modern civilization. The eradication of smallpox is one of the most important contribution toward for human and best examples of how vaccination stopped a deadly disease and saved millions of lives [4]. A vaccine is a complex biological product that can be used to safely induce an immune response that confers protection against infection and disease on subsequent exposure to a pathogen.

Vaccine adjuvants usually improve the vaccine response by stimulating the innate immune system, which provides for the rapid first line of defense against infection. Regardless of whether the vaccine is made up of the antigen itself, this weakened version will not cause the disease in the person receiving the vaccine, but it will prompt their immune system to respond much as it would have on its first reaction to the actual pathogen. To achieve this, vaccines are made from pathogenic viruses and bacteria that have been rendered less virulent by reducing their virulence. An essential component of most vaccines is one or more protein antigens that elicit an immune response that provides protection.

2.2 Vaccines induce antibodies

Vaccination response can be understood as a measure of integrated immune function, elicited by antigen exposure and measured by antibody titer and cell-mediated response [4]. The adaptive immune response is mediated by B cells that produce antibodies (humoral immunity) and by T cells (cellular immunity). All vaccines in routine use are thought to mainly confer protection through the induction of antibodies. Immune responses to antigens may be categorized as primary or secondary responses (**Figure 1**). After vaccination, B-lymphocytes detect the antigens and respond as if a real infectious agent has invaded the body, proliferating to form identical cells that can respond to the vaccine antigens. This response from immune system, generated by the B lymphocytes, is known as the primary response.

After initial antigen exposure, it takes several days for this adaptive response to become active. After the first exposure to a pathogen, immune activity rises and then levels off and declines. Since the initial immune response is slow, it does not prevent disease. Antibody levels in the circulation wane after primary vaccination, often to a level below that required for protection. During subsequent exposures to

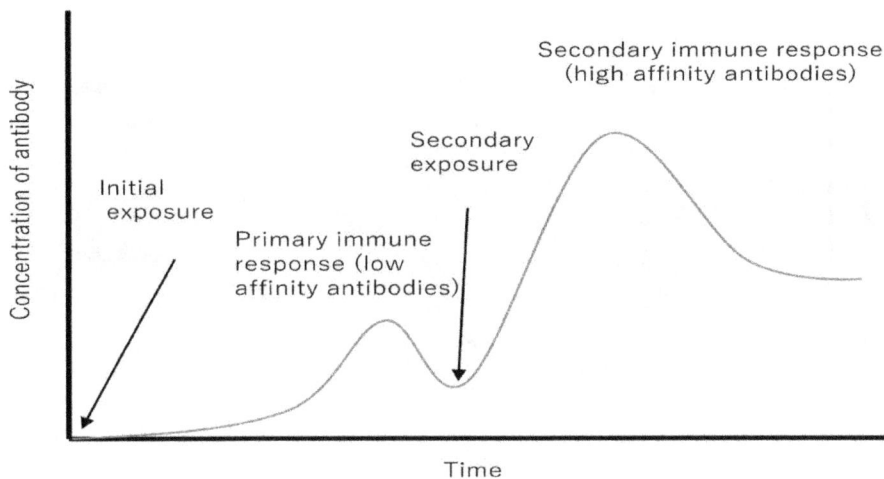

Figure 1.
During the primary response, naive B cell differentiation and antibody production occur several days after antigen encounter (initial exposure). In contrast, following secondary antigenic exposure, B cells expand with a shortened lag phase and produce larger quantities of antibodies. The difference between the primary and secondary exposures is the presence of memory B cells and pre-existing antigen-specific antibody. Memory cells differentiate into antibody-secreting plasma cells that output a greater amount of antibody for a longer period of time.

the same pathogen, the immune system can respond rapidly, and activity reaches higher levels. The secondary immune responses can usually prevent disease. In encountering a pathogen, the immune system of an individual who has been vaccinated against that specific pathogen is able to mount a protective immune response more rapidly and more robustly. Immune memory is important feature of vaccine-induced protection. Memory of the infection is reinforced, and long-lived antibodies remain in circulation. It takes several days to build to maximum intensity, and the antibody concentration in the blood peaks at about 14 days [5].

Some infections, such as chickenpox, induce a life-long memory of infection [6]. Other infections, such as influenza, vary from season to season to such an extent that even an adult is unable to adapt [7]. Seasonal influenza A and B viruses are constantly evolving in nature, often resulting in antigenic change or "drift" [8]. The composition of influenza vaccines is updated annually to keep pace with antigenic drift [9]. Whether immune memory can protect against a future pathogen encounter depends on the incubation time of the infection, the quality of the memory response and the level of antibodies induced by memory B cells.

3. Exercise and antibody response

3.1 Exercise induces antibody production

The immune response is a complex mechanism, but it is important to understand this in order to consider the possibility that it may be modulated by exercise. Liu and Wang [10] examined exercise-induced blood antibody levels in mice. They examined the plasma antibody levels of mice after infected with *Salmonella typhi*. The results showed that the antibody titer of the exercising mice was significantly higher (2.76 times higher) than that of the non-exercising group during the experiment. The antibody titer of the exercising mice was 2.76 times higher than that of the control group (**Figure 2**). Authors observed that after the initial immunization,

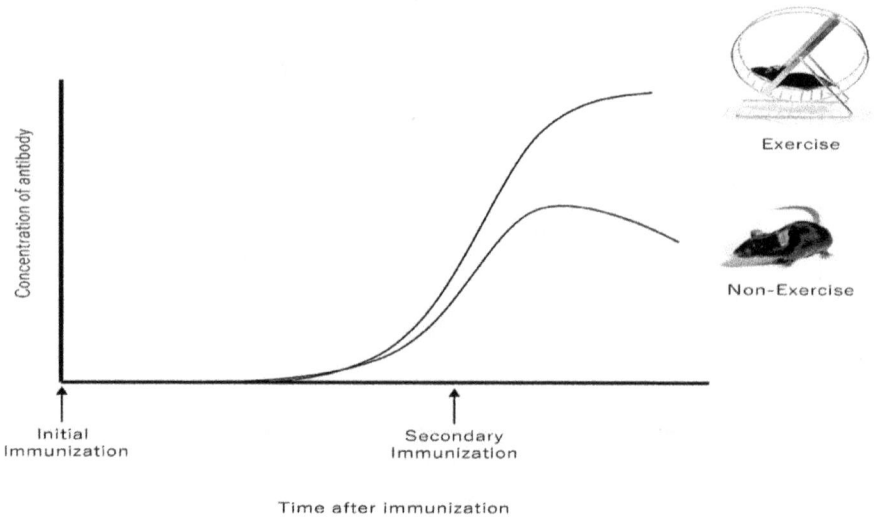

Figure 2.
Effects of physical training on the murine immunological response. Serum antibody levels in active (exercise) and control (non-exercise) mice (modified from Douglass [11]).

a primary antibody response occurred. After booster immunization, the antibody levels increased and then remained high in the blood. This result means that maintenance of long-term antibody responses is critical for protective immunity against many pathogens. After this study, effects of exercise on secondary antibody responses have been tested in young mice [12] and rats [13]. Several studies have been conducted on exercise-induced elevation of blood antibody, focusing mainly on the secondary antibody response. Exercise immunologists were intrigued by the dramatic changes in secondary antibody responses in exercising mice.

Moderate exercise, such as voluntary wheel running exercise [12] or exercise (8–15 m/minutes) on a treadmill [14], has been shown to have a marked effect on the increase in antibody levels after booster immunization. These findings have proved to be valuable information that prompted exercise immunologists to investigate. Thus, since the early days, the effect of exercise on the increase in blood antibody concentration after booster immunization has been investigated. Moderate exercise may be a powerful adjuvant to vaccination.

The rodent model is known to affect the antibody response after booster immunization. In mice, the primary IgG response to the antigen was not enhanced, but in mice subjected to exercise, the IgG response was enhanced after booster immunization [12]. Subsequent reports also showed that antibody production in exercising mice was enhanced after booster immunization [10, 11, 15]. It is unclear why moderate exercise affects the antibody response after booster immunization, while it does not affect antibody production after initial immunization.

3.2 Effect of moderate intensity physical exercise on antibody response

Both moderate and vigorous-intensity physical activity improve antibody response. Moderate-intensity physical exercises stimulate cellular immunity, while prolonged or high-intensity practices without appropriate rest can trigger decreased cellular immunity, increasing the propensity for infectious diseases [14]. Furthermore, acute and intensive exercise, more common among athletes such as marathon runners, can lead to transient immunodepression [16]. According to the International Society for Exercise and Immunology (ISEI), the immunological decrease occurs after the practice of prolonged physical exercise, that is, after 90 minutes of moderate- to high-intensity physical activity [17].

In contrast, moderate intensity physical activity is responsible for providing an increase in the immune response. As an example, elderly women participating in a moderate intensity physical exercise program aerobic exercises were performed between 60 and 70% of VO_2max, involving at least 30 minutes of exercises in step, jump coordination, and rhythmic movements sometimes dance for at least 12 months (1 hour exercise sessions 4 times a week) produced higher levels of anti-influenza (IgM and IgG) antibodies compared to sedentary women [18]. Another study showed that elderly subjects who performed physical exercised at 65–75% heart rate reserve (HRR), 25–30 minutes, 3 days per week, for 10 months also confirmed that the exercise increased the concentration of antibodies against the influenza vaccine [19].

Moderate-intensity exercise was also effective in increasing the effectiveness of the pneumococcal vaccine. When young adults immunized with pneumococcal vaccine were given 15 minutes of moderate exercise (30 seconds of exercise followed by 30 seconds of rest), they showed higher antibody production than those who did not exercise [20]. Thus, moderate physical exercise helps our body trigger the antigen specific antibody response to effectively.

3.3 Effect of exercise term on secondary antibody response

How long exercise does it take to be effective exercise induce secondary antibody response? Moderate exercise conducted over a 2- to 8-week period enhances secondary antibody response and is mediated. Kapasi et al. compared different duration of moderate exercise training on antibody immune responses in young mice [21]. Female C57BL/6 mice were randomized into 2 to 8-week exercise training or sedentary control group. Mice with 2 weeks of exercise showed a significant increase in antibodies after the additional immunization, comparable to mice with 8 weeks of exercise. Studies on the effect of moderate exercise on increased antibody levels after additional immunization will be reviewed in a later section. A moderate exercise program of 2 weeks may be sufficient to improve secondary antibody production. The author proposed that may be a useful strategy to enhance antibody response to vaccinations in humans.

3.4 Effect of exercise on antigen-specific antibody producing B cell and T cell

Factors responsible for the enhance antibody level after booster immunization have been investigated in detail by Suzuki and Tagami [12]. They examined the effect of exercise on antigen-specific IgG-producing cells in splenic lymphocytes by Enzyme-Linked ImmunoSpot [22]. The antigen-specific IgG-producing cells were significantly higher in the exercising group than in the sedentary group. Authors proposed that effects of voluntary wheel-running exercise on the number of cells which produce tetanus toxoid (TT)-specific IgG producing cells (**Figure 3**). Voluntary exercise of moderate intensity (60–70% VO_2max) increases the immune response of CD4+ T cells in healthy mice after vaccination [23]. Rogers et al. reported that exercised C57BL/6 mice with OVA intranasally immunization, and significantly increased CD4+ T cells (collected in spleen, mesentery lymph nodes, and Peyer's patches), TNF-α OVA-specific, and IL-5 were significantly increased. These reports suggest that exercise also effect on B cell and T cell responses.

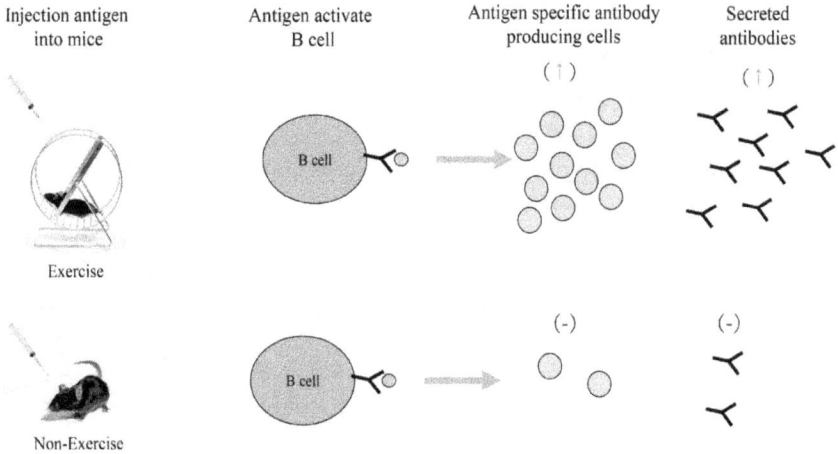

Figure 3.
Effects of physical training on the murine immunological response. Antigen specific antibody levels in active (exercise) and control (non-exercise) mice (modified from Suzuki and Tagami [12]).

3.5 Effects of exercise on endogenous opioids

Beta-endorphin, an opioid peptide is released into the blood after moderate exercise [24]. However, this phenomenon varies among individuals. Furthermore, endorphin levels in the blood are maintained for 15–60 minutes after exercise [25]. The role of endogenous opioids in exercise-induced increases in secondary antibody concentrations is unknown [12]. It has been suggested that endogenous opioids are involved in the increase in exercise-induced secondary antibody concentrations. Endogenous opioids have been implicated in exercise-induced increases in secondary antibody concentrations. Enkephalins were first observed in the brain and endocrine system. Both endorphins and enkephalins are important regulators of pain. Endorphins have been implicated in immune function [13], pain relief [26], and response to exercise [27–29]. The role of endogenous opioids in modulating exercise-induced increases in secondary antibody concentrations, especially at the cellular level, needs to be elucidated.

Kapasi et al. [30] initially immunized mice with antigens and administered placebo or an opioid antagonist (naltrexone), while untreated mice received no intervention. The mice were then subjected to exercising for 8 weeks, followed by booster immunization. After the booster immunization, the antibody levels increased in the exercising mice. On the other hand, there was no increase in antibody levels in the mice that received the antagonist. The increase in antibody concentration by endogenous opioids was dose-dependent of intravenous injection [31]. The production of antibodies occurs as a result of the interaction of antigens retained on follicular dendritic cells with B and Th lymphocytes [32, 33].

The mechanism of exercise-induced antibody concentration is activated by the binding of opioids to specific receptors on B cells and T cells [33]. Endogenous opioids also affect the antibody response through receptors on Th (CD4+) cells and by stimulating proliferation [34, 35]. These cascades are the result of induced IL-4 production, and IL-4 increases the viability of splenic B cells [36]. Further research is needed to determine if the effect of exercise is due to increased antibody levels.

3.6 Effects of exercise on IgG half-life

The mechanism that induces exercise-induced increases in blood antibody concentrations is related to the half-life of IgG [12]. The clearance rate of IgG in blood has been found to be highly dependent on its concentration in plasma [37]. The half-life of IgG in blood at physiological concentrations is about 10-fold longer than at IgG higher concentrations [37]. IgG proteins are endocytosed [38]. IgG is induced at the cell surface and released into plasma or interstitial fluid. FcRn regulates IgG epithelial transport and recycling. FcRn binds to IgG in a pH-dependent manner binding to IgG [39, 40]. In an acidic environment, IgG binds strongly to FcRn; the IgG-FcRn complex is transported by lysosomes to the cell surface where it fuses with the cell membrane [41]. At physiological pH, the FcRn receptor has little affinity for IgG. When sorting vesicles fuse to the plasma membrane, IgG dissociates from the receptor and is rapidly released into the extracellular fluid. Clearance of IgG is increased approximately 10-fold, and the efficiency of IgG recycling is over 90% in wild-type animals expressing FcRn [40].

The effect of exercising mice on IgG clearance has been reported [42]. The clearance of IgG with exercise has been reported by Suzuki and Tagami [12]. They investigated for factors that would reduce the clearance of non-specific ^{125}I-IgG in the blood after booster immunization (**Figure 4**). High blood antibody levels may result in low clearance of antibodies. The reason for the low clearance of ^{125}I-IgG in the blood of exercising mice has not yet been elucidated; the homeostasis mechanism of IgG depends on the Fc region of IgG [42]. A possible cause of the decreased clearance is the FcRn receptor, which is expressed in the vascular endothelium in mice and has an IgG protective function [43].

The FcRn molecule is dependent on dimerization with β2-microglobulin (β2m) [43]. In β2m-deficient mice, a shortening of the half-life of IgG occurs and homeostatic IgG levels are reduced [44]. Suzuki et al. reported on the effects of intraperitoneal immunization of mice with TT to induce primary and secondary antibody responses, protection from IgG catabolism in the liver and β2m expression. The authors reported an exercise-induced increase in blood antibody concentrations and a prolonged half-life of antigen-specific IgG in active mice [45]. Exercising mice had higher levels of radiolabeled IgG in the liver. This phenomenon was also confirmed by immunohistochemical analysis. The expression of the β2m gene was up regulated in the liver of exercised mice. There was a significant correlation between the amount of IgG accumulated in the liver and the concentration of IgG in the blood. There was also a significant correlation between total liver IgG and liver β2m.

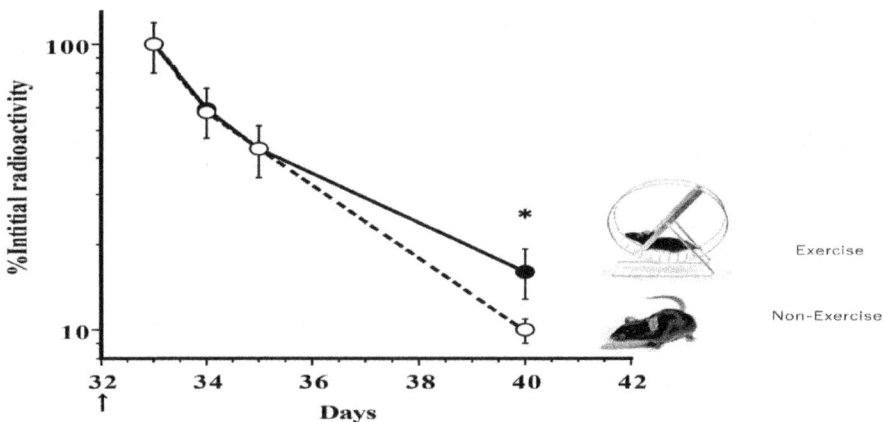

Figure 4.
Clearance of radiolabeled IgG from exercise (black circles) and non-exercise (white circles) mice (modified from Suzuki and Tagami [12]).

4. Exercise and vaccine in older population

4.1 Immune senescence in old adults

Preventive medicine is the most effective and feasible strategy to protect health in old subjects and vaccination against the most common infectious diseases is the most indicated approach. Most currently used vaccines are less immunogenic and effective in the elderly compared to younger adults [46]. This is due to several factors, including immune aging and a different immune response in children and young adults than in older adults with a history of infection. Almost all vaccines are specifically designed for children and young adults. The mechanisms of immune senescence are multiple but seem to be driven largely by changes in T cell-mediated immunity. There are fewer antigen-naïve T cells in the peripheral blood of aged individuals than there are in younger individuals [47].

Aging is a natural process and is described as "immune senescence." Immune senescence is associated with a decline in the immune system [48]. An important sign of immune senescence is the decline in immune function. Decline in immune function can lead to the development of opportunistic infections [49]. Immune senescence also results in a reduced vaccine response [50], leading to an increased incidence of infectious diseases. Improving immune function is expected to reduce the incidence of infections in the elderly and have beneficial effects in maintaining health. Moderate exercise has been used as an intervention to combat the aging of the immune system.

Aging is associated with declines in humoral and cellular immunity [51], and therefore reduced immune function. The decline in immunity due to aging is more pronounced in acquired immunity than in natural immunity. The capacity of this acquired immunity peaks around the age of 20s and declines to about half of that in the 40s. The main cause of immune senescence thought to be the change in the quality of T cells due to the decline in the function of the thymus gland.

The age-related decline in the function of major cells that take part in the antibody response are reflected by the secondary antibody response [52]. Kapasi et al. focused on age-related changes in immune function and the effects of exercise, and their study clearly showed that older mice exhibited a secondary antibody response similar to that seen in young control mice after exercise [15]. Thus, intense exercise exerts positive effects on the secondary antibody response in older animals. This, exercise-induced attenuation of immune senescence might help to improve immune responses to vaccination. Therefore, the health of the elderly is closely related to the maintenance of immune function. Therefore, the health of the elderly is closely related to the maintenance of the immune system. Thus, exercise has a positive effect on the secondary antibody response in older animals. This suggests that exercise in older animals may contribute to the immune response to vaccination. Hence, the health of the elderly is closely related to the maintenance of immune function.

4.2 Effects of exercise on vaccine response in older adult

Recently, several strategies have been tested to improve the efficacy of a vaccine in older adults. Regular exercise has been associated with enhanced vaccination responses [51, 52]. In contrast, acute exercise had no detrimental effect on vaccination response in healthy older adults [53]. Exercise-induced elevation of antibody concentrations is a tool against infectious diseases. The use of exercise-induced elevation of antibody concentrations to combat infectious diseases in humans has been investigated, and positive results have been observed [19]. Moderate aerobic exercise in the elderly [19] and resistance exercise have been shown to enhance the immune response to influenza vaccination [54, 55]. In addition, several cross-sectional studies have found that physically fit [56] and active older adults [57] have higher antibody responses to booster immunization. Shuler et al. [58] examined antibody titers to influenza vaccination in the elderly. There was no association between physical activity level and the degree of antibody concentration. However, this study did not measure antigen-specific antibodies.

Cross-sectional studies in elderly populations have all reported increased antibody concentrations after booster immunization in participants with high physical fitness [56, 59] or physical activity [57, 58, 60]. This effect of exercise-induced increases in antibody concentrations after vaccination in an elderly population was exemplified by Smith et al. [60]. They compared antibody levels in exercise-induced antibody production by immunizing with keyhole limpet hemocyanin (KLH). The results showed that antibody responses and cell-mediated responses to KLH were stronger in active elderly men than in sedentary elderly men.

Woods et al. demonstrated that 10 months of aerobic exercise (60–70% maximal oxygen uptake, 45–60 minutes, 3 times per week) in previously sedentary elderly subjects resulted in increased blood antibody levels compared to participants who only participated in flexibility training during the same period [61]. Elevated antibody concentrations to antigens have also been observed after chronic exercise; after KLH vaccination, IgG1 and IgM concentrations were higher in participants who completed a 10-month cardiovascular training program than in control participants [62]. Previously reported studies supported the hypothesis that regular exercise improves immune function in the elderly. This is reflected in the increased concentration of antigen-specific antibodies after vaccination, especially in the elderly.

5. Conclusion

In this review, we presented to improve vaccine response through moderate exercise. Currently available evidence shows that moderate exercise impacts upon the secondary antibody response but not the primary response. These effects are mediated by a diverse range of factors, including the functions of antibody producing cells, proliferation of CD4+ T cells, endogenous opioids, and IgG half-life. Over 2-week period enhances secondary antibody response is mediated. Most exercise studies have focused on antibody production, with more work required in this area. Almost all studies have investigated the effects of moderate exercise on immune function. This will provide insight into vaccination that is improved by exercise. The incorporation of molecular biological methods into the field of exercise immunology should improve our understanding of the activation of cells involved in exercise-induced increases in antibody concentrations after booster immunization.

Acknowledgements

I would like to express my heartfelt gratitude to my advisor, Dr. Kazumi Tagami, for her guidance and encouragement throughout my PhD. Dr. Tagami was always happy with me when I succeeded and worked with me to find the cause of my failures. This study was supported in part by grants from the Daiwa Securities Health Foundation and a Grant-in-Aid for Scientific Research from the Ministry of Education, Science, Sports and Culture of Japan (No. 16650154 to KT). This paper is a revised and enhanced version of the authors' recent publication [12, 45].

Abbreviations

CD	cluster of differentiation
TNF	tumor necrosis factor
OVA	ovalbumin
KLH	keyhole limpet

Author details

Kotaro Suzuki
Faculty of Health and Sports Sciences, University of Tsukuba, tsukubai-shi, ibaraki-ken, Japan

*Address all correspondence to: koutarousuzuki55@gmail.com

IntechOpen

References

[1] Havers F, Sokolow L, Shay DK, Farley MM, Monroe M, Meek J, et al. Case-control study of vaccine effectiveness in preventing laboratory-confirmed influenza hospitalizations in older adults, United States, 2010-2011. Clinical Infectious Diseases. 2016;**63**(10):1304-1311

[2] Campbell JP, Turner JE. Debunking the myth of exercise-induced immune suppression: Redefining the impact of exercise on immunological health across the lifespan. Frontiers in Immunology. 2018;**9**:648

[3] Khammassi M, Ouerghi N, Said M, Feki M, Khammassi Y, Pereira B, et al. Continuous moderate-intensity but not high-intensity interval training improves immune function biomarkers in healthy young men. Journal of Strength and Conditioning Research. 2020;**34**(1):249-256

[4] WHO. Summary report on first, second and third generation smallpox vaccines. WHO Report. 2013. p. 33

[5] Stratton K, Ford A, Rusch E, Clayton EW. Adverse Effects of Vaccines: Evidence and Causality. Washington DC: The National Academies Press; 2012. pp. 1-894

[6] Gershon AA, Breuer J, Cohen JI, Cohrs RJ, Gershon MD, Gilden D, et al. Varicella zoster virus infection. Nature Reviews. Disease Primers. 2015;**1**:1-19

[7] WHO. Evaluation of Influenza Vaccine Effectiveness: A Guide to the Design and Interpretation of Observational Studies. Geneva, Switzerland: World Health Organization; 2017

[8] Petrova VN, Russell CA. The evolution of seasonal influenza viruses. Nature Reviews. Microbiology. 2018;**16**(1):47-60

[9] Paules C, Subbarao K. Influenza. The Lancet. 2017;**390**(10095):697-708

[10] Liu YG, Wang SY. The enhancing effect of exercise on the production of antibody to *Salmonella typhi* in mice. Immunology Letters. 1987;**14**(2):117-120

[11] Douglass JH. The effects of physical tracing on the immunological response in mice. The Journal of Sports Medicine and Physical Fitness. 1974;**14**(1):48-54

[12] Suzuki K, Tagami K. Voluntary wheel-running exercise enhances antigen-specific antibody-producing splenic B cell response and prolongs IgG half-life in the blood. European Journal of Applied Physiology. 2005;**94**(5-6):514-519

[13] Kaufman JC, Harris TJ, Higgins J, Maisel AS. Exercise-induced enhancement of immune function in the rat. Circulation. 1994;**90**(1):525-532

[14] Pedersen BK, Hoffman-Goetz L. Exercise and the immune system: Regulation, integration, and adaptation. Physiological Reviews. 2000;**80**(3):1055-1081

[15] Kapasi ZF, Catlin PA, Joyner DR, Lewis ML, Schwartz AL, Townsend EL. The effects of intense physical exercise on secondary antibody response in young and old mice. Physical Therapy. 2000;**80**(11):1076-1086

[16] Cox AJ, Gleeson M, Pyne DB, Callister R, Hopkins WG, Fricker PA. Clinical and laboratory evaluation of upper respiratory symptoms in elite athletes. Clinical Journal of Sport Medicine. 2008;**18**(5):438-445

[17] Walsh NP, Gleeson M, Pyne DB, Nieman DC, Dhabhar FS, Shephard RJ, et al. Position statement. Part two: Maintaining immune health. Exercise Immunology Review. 2011;**17**:64-103

[18] Bachi ALL, Suguri VM, Ramos LR, Mariano M, Vaisberg M, Lopes JD. Increased production of autoantibodies and specific antibodies in response to influenza virus vaccination in physically active older individuals. Results in Immunology. 2013;**3**:10-16

[19] Kohut ML, Arntson BA, Lee W, Rozeboom K, Yoon K-J, Cunnick JE, et al. Moderate exercise improves antibody response to influenza immunization in older adults. Vaccine. 2004;**22**(17-18):2298-2306

[20] Edwards KM, Pung MA, Tomfohr LM, Ziegler MG, Campbell JP, Drayson MT, et al. Acute exercise enhancement of pneumococcal vaccination response: A randomised controlled trial of weaker and stronger immune response. Vaccine. 2012;**30**(45):6389-6395

[21] Kapasi ZF, Catlin PA, Adams MA, Glass EG, McDonald BW, Nancarrow AC. Effect of duration of a moderate exercise program on primary and secondary immune responses in mice. Physical Therapy. 2003;**83**(7):638-647

[22] Czerkinsky C, Andersson G, Ekre HP, Nilsson LA, Klareskog L, Ouchterlony O. Reverse ELISPOT assay for clonal analysis of cytokine production. I. Enumeration of gamma-interferon-secreting cells. Journal of Immunological Methods. 1988;**110**(1):29-36

[23] Rogers CJ, Zaharoff DA, Hance KW, Perkins SN, Hursting SD, Schlom J, et al. Exercise enhances vaccine-induced antigen-specific T cell responses. Vaccine. 2008;**26**(42):5407-5415

[24] Pilozzi A, Carro C, Huang X. Peran β-Endorphin dalam Stres, Perilaku, Peradangan Neuroin, dan Metabolisme. International Journal of Molecular Sciences. 2021;**22**(1):338-362

[25] Sforzo GA. Opioids and exercise. An update. Sports Medicine. 1989;**7**(2):109-124

[26] Sommers DK, Loots JM, Simpson SF, Meyer EC, Dettweiler A, Human JR. Circulating met-enkephalin in trained athletes during rest, exhaustive treadmill exercise and marathon running. European Journal of Clinical Pharmacology. 1990;**38**(4): 391-392

[27] Jamurtas AZ, Goldfarb AH, Chung SC, Hegde S, Marino C. Beta-endorphin infusion during exercise in rats: Blood metabolic effects. Medicine and Science in Sports and Exercise. 2000;**32**(9):1570-1575

[28] Nauli SM, Maher TJ, Pearce WJ, Ally A. Effects of opioid receptor activation on cardiovascular responses and extracellular monoamines within the rostral ventrolateral medulla during static contraction of skeletal muscle. Neuroscience Research. 2001;**41**(4):373-383

[29] Tsuchimochi H, McCord JL, Kaufman MP. Peripheral mu-opioid receptors attenuate the augmented exercise pressor reflex in rats with chronic femoral artery occlusion. American Journal of Physiology. Heart and Circulatory Physiology. 2010;**299**(2):H557-H565

[30] Kapasi ZF, Catlin PA, Beck J, Roehling T, Smith K. The role of endogenous opioids in moderate exercise training-induced enhancement of the secondary antibody response in mice. Physical Therapy. 2001;**81**(11):1801-1809

[31] Janković BD, Marić D. Enkephalins and immunity. I: In vivo suppression and potentiation of humoral immune response. Annals of the New York Academy of Sciences. 1987;**496**: 115-125

[32] Mousa SA, Zhang Q, Sitte N, Ji R, Stein C. Beta-endorphin-containing memory-cells and mu-opioid receptors undergo transport to peripheral inflamed tissue. Journal of Neuroimmunology. 2001;**115**(1-2):71-78

[33] Tew JG, Kosco MH, Burton GF, Szakal AK. Follicular dendritic cells as accessory cells. Immunological Reviews. 1990;**117**:185-211

[34] Kamphuis S, Eriksson F, Kavelaars A, Zijlstra J, van de Pol M, Kuis W, et al. Role of endogenous pro-enkephalin A-derived peptides in human T cell proliferation and monocyte IL-6 production. Journal of Neuroimmunology. 1998;**84**(1):53-60

[35] Sharp BM, Li MD, Matta SG, McAllen K, Shahabi NA. Expression of delta opioid receptors and transcripts by splenic T cells. Annals of the New York Academy of Sciences. 2000;**917**:764-770

[36] Mori M, Morris SC, Orekhova T, Marinaro M, Giannini E, Finkelman FD. IL-4 promotes the migration of circulating B cells to the spleen and increases splenic B cell survival. Journal of Immunology. 2000;**164**(11): 5704-5712

[37] Waldmann TA, Strober W. Metabolism of immunoglobulins. Progress in Allergy. 1969;**13**:1-110

[38] Jones DH, Nusbacher J, Anderson CL. Fc receptor-mediated binding and endocytosis by human mononuclear phagocytes: Monomeric IgG is not endocytosed by U937 cells and monocytes. The Journal of Cell Biology. 1985;**100**(2):558-564

[39] Brambell FW, Hemmings WA, Morris IG. A theoretical model of gamma-globulin catabolism. Nature. 1964;**203**:1352-1354

[40] Junghans RP. Finally! The Brambell receptor (FcRB). Mediator of transmission of immunity and protection from catabolism for IgG. Immunologic Research. 1997;**16**(1):29-57

[41] Qiao S-W, Kobayashi K, Johansen F-E, Sollid LM, Andersen JT, Milford E, et al. Dependence of antibody-mediated presentation of antigen on FcRn. Proceedings of the National Academy of Sciences of the United States of America. 2008;**105**(27):9337-9342

[42] Ghetie V, Hubbard JG, Kim JK, Tsen MF, Lee Y, Ward ES. Abnormally short serum half-lives of IgG in beta 2-microglobulin-deficient mice. European Journal of Immunology. 1996;**26**(3):690-696

[43] Wani MA, Haynes LD, Kim J, Bronson CL, Chaudhury C, Mohanty S, et al. Familial hypercatabolic hypoproteinemia caused by deficiency of the neonatal Fc receptor, FcRn, due to a mutant beta2-microglobulin gene. Proceedings of the National Academy of Sciences of the United States of America. 2006;**103**(13):5084-5089

[44] Christianson GJ, Brooks W, Vekasi S, Manolfi EA, Niles J, Roopenian SL, et al. Beta 2-microglobulin-deficient mice are protected from hypergammaglobu-linemia and have defective antibody responses because of increased IgG catabolism. Journal of Immunology. 1997;**159**(10):4781-4792

[45] Suzuki K, Suk PJ, Hong C, Imaizumi S, Tagami K. Exercise-induced liver beta2-microglobulin expression is related to lower IgG clearance in the blood. Brain, Behavior, and Immunity. 2007;**21**(7):946-952

[46] Jefferson T, Rivetti D, Rivetti A, Rudin M, Di Pietrantonj C, Demicheli V. Efficacy and effectiveness of influenza vaccines in elderly people: A systematic review. The Lancet. 2005;**366**(9492): 1165-1174

[47] Suchard M. Immunosenescence: Ageing of the immune system. South African Pharmaceutical Journal. 2015;**82**:28-31

[48] Gruver AL, Hudson LL, Sempowski GD. Immunosenescence of ageing. The Journal of Pathology. 2007;**211**(2):144-156

[49] Pawelec G, Adibzadeh M, Pohla H, Schaudt K. Immunosenescence: Ageing of the immune system. Immunology Today. 1995;**16**:420-422

[50] Kumar R, Burns EA. Age-related decline in immunity: Implications for vaccine responsiveness. Expert Review of Vaccines. 2008;7(4):467-479

[51] Hodes RJ. Aging and the immune system. Immunological Reviews. 1997;**160**:5-8

[52] Szakal AK, Kapasi ZF, Masuda A, Tew JG. Follicular dendritic cells in the alternative antigen transport pathway: Microenvironment, cellular events, age and retrovirus related alterations. Seminars in Immunology. 1992;**4**(4): 257-265

[53] Bohn-Goldbaum E, Pascoe A, Singh MF, Singh N, Kok J, Dwyer DE, et al. Acute exercise decreases vaccine reactions following influenza vaccination among older adults. Brain, Behavior, & Immunity - Health. 2020;**1**:100009

[54] Schuler PB, Lloyd LK, Leblanc PA, Clapp TA, Abadie BR, Collins RK. The effect of physical activity and fitness on specific antibody production in college students. The Journal of Sports Medicine and Physical Fitness. 1999;**39**(3):233-239

[55] Edwards KM, Burns VE, Allen LM, McPhee JS, Bosch JA, Carroll D, et al. Eccentric exercise as an adjuvant to influenza vaccination in humans. Brain, Behavior, and Immunity. 2007;**21**(2): 209-217

[56] Keylock KT, Lowder T, Leifheit KA, Cook M, Mariani RA, Ross K, et al. Higher antibody, but not cell-mediated, responses to vaccination in high physically fit elderly. Journal of Applied Physiology. 2007;**102**(3):1090-1098

[57] Kohut ML, Cooper MM, Nickolaus MS, Russell DR, Cunnick JE. Exercise and psychosocial factors modulate immunity to influenza vaccine in elderly individuals. The Journals of Gerontology. Series A, Biological Sciences and Medical Sciences. 2002;**57**(9):M557-M562

[58] Schuler PB, Leblanc PA, Marzilli TS. Effect of physical activity on the production of specific antibody in response to the 1998-99 influenza virus vaccine in older adults. The Journal of Sports Medicine and Physical Fitness. 2003;**43**(3):404

[59] Brydak LB, Tadeusz S, Magdalena M. Antibody response to influenza vaccination in healthy adults. Viral Immunology. 2004;**17**(4):609-615

[60] Smith TP, Kennedy SL, Fleshner M. Influence of age and physical activity on the primary in vivo antibody and T cell-mediated responses in men. Journal of Applied Physiology. 2004;**97**(2): 491-498

[61] Woods JA, Keylock KT, Lowder T, Vieira VJ, Zelkovich W, Dumich S, et al. Cardiovascular exercise training extends influenza vaccine seroprotection in sedentary older adults: The immune function intervention trial. Journal of the American Geriatrics Society. 2009;**57**(12):2183-2191

[62] Grant RW, Mariani RA, Vieira VJ, Fleshner M, Smith TP, Keylock KT, et al. Cardiovascular exercise intervention improves the primary antibody response to keyhole limpet hemocyanin (KLH) in previously sedentary older adults. Brain, Behavior, and Immunity. 2008;**22**(6):923-932

www.ingramcontent.com/pod-product-compliance
Lightning Source LLC
Chambersburg PA
CBHW081539190326
41458CB00015B/5590